PROCEEDINGS OF THE FOURTH CONFERENCE ON

ORIGINS OF LIFE

CHEMISTRY AND RADIOASTRONOMY

PROCEEDINGS OF THE FOURTH CONFERENCE ON

⌐ORIGINS OF LIFE
CHEMISTRY AND RADIOASTRONOMY⌐, *1973*

Edited by

LYNN MARGULIS

SPRINGER-VERLAG   NEW YORK · HEIDELBERG · BERLIN
1973

Origins of Life, Volume IV

This Conference was organized and conducted by The Interdisciplinary Communications Program of the Smithsonian Institution and was supported by the National Aeronautics and Space Administration under contract number NSR-09-015-044. It was held at Belmont Conference Center in Elkridge, Maryland, from April 13-16, 1971.

ISBN: 0-387-06066-9   Springer-Verlag   New York · Heidelberg · Berlin
ISBN: 3-540-06066-9   Springer-Verlag   Berlin · Heidelberg · New York

Printed in the United States of America

TABLE OF CONTENTS

v

235209

LIST OF FIGURES

LIST OF TABLES

This is an edited record of the dialogue of eminent scientists attending the fourth conference in the series of the origins of life, supported by a grant from the Biosciences Program of the National Aeronautics and Space Administration. The first conference at Princeton, 1967, was held under the direction of Dr. Frank Fremont-Smith at the time when the Interdisciplinary Communications Program (ICP) was associated with the New York Academy of Sciences.

In 1968, ICP was integrated into the Office of the Assistant Secretary for Science of the Smithsonian Institution, and the entire operation was set up in Washington, D. C. The second conference, also in Princeton, was held in 1968 and the third was in Santa Ynez, California in 1970. (See Margulis, ed. 1970, 1971 and 1972 for previously published proceedings.) The fourth conference was held at the Belmont Conference Center in Elkridge, Maryland, April 13-16, 1971.

The proceedings are recorded and edited by the Interdisciplinary Communications Associates, Inc. (ICA, a nonprofit foundation), for ICP. Dr. Lynn Margulis, who served as Scientific Editor for the entire series, has succeeded admirably in carrying out a difficult and usually inadequately appreciated task. She was ably assisted in her offices by Barbara Miranda and Harriet Eklund, the ICP Staff Editor for this series. I am grateful to all.

ICA was formed to encourage effective interchange and interaction among the various scientific and social disciplines and to aid in the solutions of scientific and social problems. Currently its primary concern is with assisting ICP.

M. C. Shelesnyak, Ph.D.

Director, Interdisciplinary
   Communications Program
Smithsonian Institution
Washington, D. C.

This is the last of a series of informal conferences on the *Origins of Life*. The first, in May 1967, concerned prebiotic organic syntheses experiments and the early fossil record (Margulis, ed. 1970). Most of us left the conference with a deeper realization of the antiquity of terrestrial life, having seen evidence for microbes in the very earliest sediments of South Africa. With respect to the other topics discussed at that conference, research findings in "prebiotic syntheses" were exciting and many new biomonomers and polymers have been synthesized in better understood ways in the laboratory under plausibly early earth conditions. Yet, the experiments, taken together, still seemed hopelessly far away from producing the simplest cell, that is, the minimal replicating and evolving system with its membrane-bound, coupled nucleic acid-protein synthesizing system.

Since life turned out to be so ancient, it was thought that perhaps conditions that led to its appearance occurred before the stabilization of the earth's crust and the onset of the last three and a half eons of the geological cycle. Thus, the second conference, also at Princeton (May 1968) brought together distinguished cosmologists, astronomers, and geologists who explored the question of the origin of the earth (Margulis, ed. 1971). This event, like the origin of life, also seems shrouded in mystery and not to have occurred under conditions terribly conductive to life as we know it.

Because in 1969 there was such a rapid influx of data from our neighbors in the solar system, the third conference was postponed a year. Held in February 1970, it explored the possible solutions to the origins of life through studies of the earth's relations with the moon, Venus, and Mars (Margulis, ed. 1972). Several of the participants were principal investigators on the Apollo projects and came with important new information about the lunar rocks. Stunning photos accompanied by analyses of lunar and Martian cratered terrain were shown.

What bearing had the recent explorations of these close neighbors on the problem at hand? The moon is old and has been dead for a long time. It is very, very dry. Mars, although far more active, is cold and barren. The Martian atmosphere of CO_2 (and lack of N_2) told me (perhaps because some articulate partisans were persuasive) that not too much goes on there on Saturday nights; therefore, it would appear that the onus of proof that Mars harbors life seems to now lie with the exobiologists. I left that exciting third conference imbued with confidence in the methods of science coupled with grave doubts that reproducing, evolving and adapting life will be found anywhere in the nearby solar system. We anticipate the results of the Viking Mars project in 1976 to resolve at least some of the issues raised that year.

This fourth conference provides us with an enthusiastic and informative dialogue of this unconventional and interdisciplinary field called "the origin of life." Carbon dioxide and monoxide may be incorporable into reduced organic matter under Martian conditions of high uv fluxes (p. 4 ff). I may be wrong: Mars may indeed provide a celestial laboratory as the early solar system "control" for the earth's biological experiment. It may even house a wierd, modest biota. No matter the detailed outcome of the life detection experiments of the Viking mission both exo- and prebiotic experimenters had better heed the nearly-born "Belmont Manifesto" (p. 34) and apply those same rigorous criteria of identification that characterize other branches of organic chemistry. From this fourth conference a sense of scientific well-being emerges in part because of some very recent developments, namely, the sweeping general applicability of "plate tectonic" concepts that permits the comprehension of discrete fossil faunas and floras, rates of sedimentation and outgassing, movements of continents, and formation of deep ocean trenches with a single set of ideas (p. 170 ff). Even though the role of organic geochemistry in elucidating our evolutionary past is still equivocal (p. 124 ff), at the end of this conference we understood better the history of the earth with its mobile crust and its ever adapting ancient life. Yet, it must be admitted after this five-year attack from various angles, prejudices and disciplines, the central problem inspiring these conferences, perhaps slightly better defined, is as unsolved as ever. Did our organic matter originate in interstellar space? The infant science of radioastronomy (p. 219 ff) has produced evidence that some of the small organic molecules are there. Is the presence of HCN, H_2CO, NH_3 and so forth relevant to the origin of terrestrial life? Those seeking a sure unequivocal and complete answer to the origin of life question will be disappointed. However, there is no reason for "hard scientists" to abandon these problems as hopelessly unsolvable. For example, when greenhouse calculations jibed with considerations of minimal ammonium concentrations for the prebiotic production of a central amino acid (aspartate), we began to feel there was at least some ammonia in the early atmosphere. When it was calculated that this hypothesized atmospheric ammonia has the fortuitous property of preventing the oceans from freezing at the critical period when the sun gave off less radiation (p. 196 ff), we felt that progress is being made in the reconstruction of our elusive past.

We now recognize that if the origin of our self-replicating system occurred on the early earth it must have occurred quite quickly (millions not billions of years), probably at alkaline pH's, cool temperatures and under reducing atmospheric conditions. But we still don't know how.

I acknowledge with gratitude the expert editorial assistance of Barbara Miranda.

Lynn Margulis
Boston University
Spring 1972

FOURTH CONFERENCE ON ORIGINS OF LIFE
Belmont Conference Center, Elkridge, Maryland
April 13-16, 1971

Chairman, Dr. Stanley Miller
Co-Chairman: Dr. Leslie Orgel

PARTICIPANTS LIST

ABELSON, Dr. Philip
President, Carnegie Institution
  of Washington
1530 P St., N.W.
Washington, D.C. 20005

LEMMON, Dr. Richard
Biodynamics Laboratory
Lawrence Radiation Laboratory
University of California
Berkeley, California 94720

BADA, Dr. Jeffrey L.
Scripps Institution of Oceanography
University of California, San Diego
La Jolla, California 92037

LOHRMANN, Dr. R.
Salk Institute
10010 North Torrey Pines Road
La Jolla, California 92037

BARGHOORN, Dr. Elso
Department of Geology
Harvard University
Cambridge, Massachusetts 02139

MARGULIS, Dr. Lynn
Department of Biology
Boston University
Boston, Massachusetts 02215

BUHL, Dr. David
National Radio Astronomy Observatory
Edgemont Road
Charlottesville, Virginia 22901

MILLER, Dr. Stanley L.
Department of Chemistry
University of California, San Diego
La Jolla, California 92037

FERRIS, Dr. James
Department of Chemistry
Rensselaer Polytechnical Institute
Troy, New York 19180

NAGYVARY, Dr. Joseph
Department of Chemistry
Texas A & M
College Station, Texas 77843

HUBBARD, Dr. Jerry
Jet Propulsion Laboratory
4800 Oak Grove Drive
Pasadena, California 91103

ORGEL, Dr. Leslie
Salk Institute
10010 North Torrey Pines Road
La Jolla, California 92037

HULETT, Dr. H. R.
Genetics Department
Stanford University
Stanford, California 94305

ORÓ, Dr. Juan
Department of Biophysical Sciences
University of Houston
Houston, Texas 77004

OWEN, Dr. Tobias
State University of New York
  at Stony Brook
Stony Brook, Long Island
New York 11790

PONNAMPERUMA, Dr. Cyril
Department of Chemistry
University of Maryland
College Park, Maryland 20742

SAGAN, Dr. Carl
Laboratory for Planetary Studies
Cornell University
Ithaca, New York 14850

SANCHEZ, Dr. Robert
Salk Institute
10010 North Torrey Pines Road
La Jolla, California 92307

SCHOPF, Dr. J. William
Department of Geology
University of California
Los Angeles, California 90024

SHELESNYAK, Dr. M. C.
Director, Interdisciplinary
  Communications Program
1717 Massachusetts Ave. N.W.
Washington, D.C. 20036

SIEVER, Dr. Raymond
Department of Geology
Harvard University
Cambridge, Massachusetts 02138

SOFFEN, Dr. Gerald A.
Viking Project Scientist
National Aeronautics and Space
  Administration
Langley Research Center
Hampton, Virginia 23365

USHER, Dr. David
Department of Chemistry
Cornell University
Ithaca, New York 14850

WOLMAN, Dr. Y.
Department of Organic Chemistry
Hebrew University of Jerusalem
Jerusalem, Israel

YOUNG, Dr. Richard
National Aeronautics and Space
  Administration
400 Maryland Avenue, S. W.
Washington, D.C. 20546

PROCEEDINGS OF THE FOURTH CONFERENCE ON

ORIGINS OF LIFE

CHEMISTRY AND RADIOASTRONOMY

The Fourth Conference on Origins of Life of the Interdisciplinary Communications Program, Smithsonian Institution, held at the Belmont Conference Center, Elkridge, Maryland, convened at eight twenty-five o'clock, Dr. Stanley Miller, Chairman of the Conference, presiding.

MILLER: Welcome to the fourth meeting in this series on the origin of life. These conferences are unique so I thought Dr. Shelesnyak, who plans them, might describe how they work for those who haven't been here before.

SHELESNYAK: Thank you.

Had this been Friday night, I could have maintained a fine Jewish tradition by answering: Why is this night different than all other nights? [Laughter]

First, I want to welcome those who have been at these meetings before and the newcomers. The basic object of these conferences is the attempt to give those concerned with a common problem but from different points of view, an opportunity to communicate with one another. We want to break language and idea barriers, to cement old friendships and to make new ones.

Some ground rules should be spelled out. No one must give a speech or listen to a speech. Please do not sit through statements you don't understand: this is a dialogue, a discourse. Although we must let people finish their sentences, we have no formalities, and want to encourage interaction and exchange.

Stan [Miller] and I have been battling about this particular meeting. I feel that there is a critical mass of about fourteen to fifteen people, beyond which continued exchange is difficult. If someone has something important to say and is one of those rare modest, shy people, the Chairman may know he has something to say but there isn't time to include him.

I would be very happy to have an experiment showing that a critical mass of twenty-five can interact effectively. My main point is to invite you to participate in the discussion on a continuing basis.

The other usual pattern here is for a series of self-introductions. Before my time, in the program I inherited these fell into a pattern of long-winded redundancy. It might have been all right the first time somebody gave his biography in lighthearted humor but the second, third, fourth, and fifth times it was kind of deadening.

1

SAGAN: Just distribute the biographies from the last meeting. [Laughter]

SHELESNYAK: Yes, but Lynn has reduced most of those to single sentences.

We have a large group, and want to discuss substance this evening. Stan, who is the Chairman and runs the show, suggests that the biographical statements be limited to forty, plus or minus ten seconds. [Laughter] Stan, it's all yours.

MILLER: My name is Stanley Miller. I'm at the University of California, San Diego. I have been interested in the field of the origin of life for some years.

LOHRMANN: I'm Rolf Lohrmann. I'm working at the Salk Institute, San Diego, and I have spent the last five years in the field of organic chemistry.

BUHL: I'm David Buhl. I'm from the National Radio Astronomy Observatory, colloquially known as Green Bank (West Virginia). I'm in the field of astronomy and interstellar molecules, and I have been interested in the origin of life for about two months. [Laughter]

BARGHOORN: Elso Barghoorn. I'm from Harvard. I notice that you have me down as Department of Geology. Actually I'm both geology and biology, primarily biology.

I have been interested for twenty years — fifteen or eighteen at least — in this problem from the standpoint of how far back you can find tangible evidence of life.

FERRIS: Jim Ferris. I'm an organic chemist at Rensselaer Polytechnic Institute, concerned with pathways of prebiotic synthesis.

SAGAN: Carl Sagan, Cornell University.

MILLER: You are interested in everything! [Laughter]

LEMMON: Carl, you have thirty-seven seconds left. [Laughter]

SAGAN: I'll cash in my chips later. [Laughter]

BADA: Jeffrey Bada. I'm at the Scripps Institution of Oceanography, and I'm an organic chemist. I work on the stability of organic compounds in the geological environment. I became interested in the origin of life when I worked with Professor Miller on my doctoral degree.

HUBBARD: Jerry Hubbard from the Bioscience Section at JPL, Jet Propulsion Laboratory, working on the Mars experiments for about the past four years. In the past year I've become interested in the possibility of abiogenic synthesis on Mars.

LEMMON: I'm Dick Lemmon, a radiation chemist, from the University of California at Berkeley. I have been interested in the effects of high energy radiation on organic compounds relevant to the origin of life for some ten years.

NAGYVARY: I'm Joseph Nagyvary from Texas A & M, an organic chemist. I have been working on nucleotide analogs for the past six years. I have been interested in prebiotic chemistry since I had dinner with John Oró about two years ago.

SCHOPF: I'm Bill Schopf from the University of California at Los Angeles. I'm primarily interested in the Precambrian evolution of biological systems. My recent work has concerned the nature of late Precambrian organisms and the evolution of blue green algae.

YOUNG: I'm Dick Young, from NASA, National Headquarters. I'm with the exobiology program which is primarily concerned with extraterrestrial life and the origin of life.

WOLMAN: I'm Yehetzkial Wolman, from Hebrew University, spending my sabbatical year in the Department of Chemistry in the University of California at La Jolla. I'm a peptide chemist, interested in structure, function and relationships of peptides and polypeptides. I became interested in the chemistry of the origin of life a few years ago.

SIEVER: I'm Ray Siever, from Harvard. I'm a geologist. I first got interested in the origin of life when I saw some of Elso Barghoorn's ancient fossils about twelve or fifteen years ago.

USHER: I'm Dave Usher, Cornell University, Chemistry Department. I call myself a bio-organic chemist. I'm largely interested in organic phosphates and enzymes.

SANCHEZ: I'm Robert Sanchez, of the Salk Institute. I'm an organic chemist. During the past five years I have been primarily concerned with the synthesis of biomonomers; in particular, nucleosides.

SOFFEN: I'm Gerry Soffen, a biologist interested in exobiology and the origin of life for about ten years. I find I'm the boy who put his finger in the dike, and there is nobody to let me out. As Viking Project scientist, I live in a sea of engineers. We're trying to land a payload on Mars in '75 which is ultimately connected with the objectives of this conference.

HULETT: I'm Russ Hulett, from Stanford University. I think you have all seen my paper (Hulett, 1969) that was enclosed with your files. I think I've said enough! [Laughter] (See p. 77)

MARGULIS: I'm Lynn Margulis, from Boston University. My training is primarily in genetics. I have been here at this conference three times. Everything you have to say is taken down by this gentleman [indicating stenographer], and I have to see it afterwards, an impossible job! [Laughter] I'm really interested in the early evolution of life; that is, once you boys have made it chemically, what happened after that, where we have rocks, and biology to give us explicit evidence.

ORGEL: I'm Leslie Orgel. I'm with the Salk Institute. I'm a theoretical chemist — and I have been interested in the origins of life for about six or seven years.

SHELESNYAK: And I'm Shelly Shelesnyak. I run the Interdisciplinary Communications Program, something I took on about four years ago. Most of my professional career outside the government and the military was in what are called biodynamics reproduction, biology of reproduction. I'm concerned with embryonic origins: origins at a very much later stage than anybody in this group, but I'm still interested.

IDENTIFICATION OF PREBIOTIC COMPOUNDS AND THE CONTAMINATION PROBLEM

MILLER: The chairman for the evening is supposed to be Cyril Ponnamperuma, but he won't be here until tomorrow morning. I'm the discussion initiator so I'll take on both tasks.

It's assumed that life arose on the earth. We don't know the whole story by any means. In attempting to understand what primitive earth conditions were like, it seems that the organic chemistry can be as important as anything in one's considerations. An example is the general argument: The fact that organic compounds can not be made under oxidizing conditions, but can be made under reducing conditions, implies that the early earth had reducing conditions. I don't want to enter into the validity and the intricacies of this argument, but to bring out the point that organic chemistry is usually more restrictive than any geological argument in narrowing down the various choices. I expect that you will see illustrations of this during this conference.

I thought we'd begin by discussing new prebiotic syntheses. Rather than go through any that I or others have been playing with in the labs, we should hear the latest and most exciting results of experiments that have been done by Dr. Hubbard, who is here, and by Dr. Norman Horowitz and Dr. Jim Hardy (both from the Jet Propulsion Laboratory), who are not able to be here.

Norm Horowitz couldn't be here, because of his teaching load at Cal Tech, but we do have the pleasure of Dr. Hubbard: would you be willing to discuss your carbon monoxide results?

HUBBARD: I became involved in abiogenic synthesis in a rather back-handed way, which may account for what seems to be the unusual approach we took. I have been associated with Norm Horowitz and George Hobby in the development of one of the life detection experiments.

The experiment attempts to measure biology in the Martian soil under conditions as close to Martian ambient as possible. We enrich the natural

Martian atmosphere with small quantities of C^{14}-carbon dioxide, and C^{14}-carbon monoxide, and look for biological incorporation into organic materials in the soil.

There are numerous control experiments in the course of the development of a flight instrument. The data I will be discussing arose out of an observation in one of these tests. When a sample of sterilized soil under an atmosphere of ^{14}C-carbon monoxide, ^{12}C-carbon dioxide, and water vapor is irradiated with an ultraviolet source which grossly similates the spectral quality and intensity reaching the Martian surface, we observe the formation of organic compounds. For most experiments we use a high pressure xenon lamp as the UV source.

LEMMON: What's in the soil?

HUBBARD: Our first experiments employed an organic soil. Initially, we were looking for an abiotic exchange of carbon into the soil organic matter, so we used a soil rich in organic matter.

We have also used ignited soils and ignited Vycor particles. The Vycor particles are highly fractured with a very high surface area, about 170 meters per gram.

SCHOPF: Is Vycor an organic polymer?

MILLER: No. It is essentially fused silica, but not quite.

HUBBARD: Routinely, when we use organic soil we dry heat sterilize the material, then equilibrate with water vapor before adding it to the quartz chamber. After equilibration with water vapor, we then pump on it several times, so we don't know exactly how much water is retained. In the case of Vycor, we ignite to 720° in air, equilibrate with water vapor, and then place in the tube, pump on it the same way. I will refer to some experiments where we reduced the level of water vapor.

The high pressure xenon source provides a fair simulation of the solar spectrum. We used 32 milliwatts per square centimeter total intensity, compared to 42 mw/cm^2 for Martian surface at zenith. Figure 1B shows the spectrum at wavelengths below 3000 Angstroms. The xenon lamp is a good simulation except for the excess of energy around 1800 Angstroms.

SAGAN: Would you please read the abscissa?

HUBBARD: It is microwatts per square centimeter per Angstrom.

SAGAN: Thank you. That's the ordinate. [Laughter]

HUBBARD: I'm sorry. Wavelength in Angstroms from 1800 through 3000Å. Also shown are calculated values for the UV reaching the outer atmosphere of Mars (Fig. 1B).

SAGAN: Is that a linear scale for the ordinate?

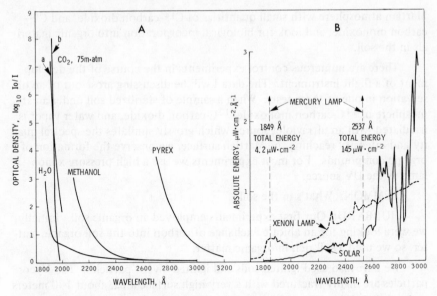

Figure 1A. UV absorption by Martian CO_2 atmosphere, 5 cm of liquid H_2O, 5 cm of methanol and 4 mm of pyrex. Optical densities of 75 m-atm of CO_2 were calculated from absorption coefficients: a = partially extrapolated data of Thompson et al. b = unpublished measured values of A. L. Lane.

Figure 1B. Solar UV spectrum reaching the outer atmosphere of Mars and the UV emission spectra of the xenon and mercury lamps.

HUBBARD: Yes.

We used the xenon source and irradiated a mixture of 0.37 per cent $C^{14}O$ and 96 per cent unlabelled CO_2, plus water vapor at about 2.8%.

SIEVER: What is the total pressure?

HUBBARD: Slightly below one atmosphere. After 140 hours' irradiation about 40 per cent of the ^{14}CO is converted to $C^{14}O_2$, and another 10 per cent of the radioactivity is converted to organic compounds extractable from our soil substratum (Fig. 2).

MARGULIS: What, exactly, is this soil?

HUBBARD: This is organic soil collected in a farming area near the Santa Anita Racetrack. [Laughter]

HULETT: Real organic!

LEMON: Would you kindly tell us a little about the rationale of these experiments and what they have to do with Mars?

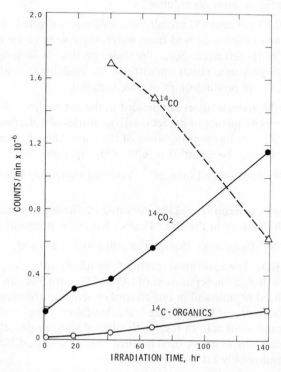

Figure 2. Conversion of labeled carbon monoxide into organic carbon as a function of irradiation time.

HUBBARD: Yes, if I may finish the description of the initial observation. This soil is an opaque material.

SIEVER: Is it all organic?

HUBBARD: No, it's a rich farm soil with about five to ten per cent organic materials.

SIEVER: Okay. Is the rest clay?

HUBBARD: I can only assume it contains clay and other things.
Using the ignited Vycor particles in this same experiment would yield about half of organics as on the organic soil.

SAGAN: Do you know because you've done this?

HUBBARD: Yes.

Figure 2 shows an experiment with 30 mg of soil substratum in a quartz tube of about 5.5 cc total gas volume.

We had 0.37 per cent CO initially and half was converted to CO_2. Assuming that the oxygen is derived from water, then we have the equivalent of 0.18 per cent of H_2 left over. Since the major product is formaldehyde,* CH_2O, we have only used about one-fifth of the available reducing potential, the hydrogen, in the production of organic material.

Apparently, reduced gases are present in the atmosphere. We have not run conclusive experiments to detect small quantities of volatiles in the gaseous atmosphere. We may be counting some of the formaldehyde in with our CO; or if it's acid, we may be counting it in the CO_2 fraction.

SOFFEN: Jerry, may I interject? You said something I don't think you meant.

When Dick [Lemmon] said incorporated by the organic material he really meant by biology in the soil. That's what the experiment is all about.

LEMMON: Thank you, though I'm still a little confused.

HUBBARD: The basic assumption of the life detection experiment is to look for the biological incorporation of CO or CO_2 into Martian soil organisms. We used sterilized organic soil in our control experiments because we were exploring the possibility of UV catalyzed, abiological exchange of CO_2 or CO into the soil organic material. In tests with labelled CO_2 we detected no exchange reactions, but when we used labelled CO, we observed radioactivity incorporated abiologically into the soil.

BARGHOORN: How was this rich soil sterilized?

HUBBARD: It was dry heat sterilized to 175° overnight.

LEMMON: I'd like to interject some information here. If you take CO_2 and a hydrocarbon with ionizing radiation you get a fatty acid — a carboxylic acid. The CO_2 is incorporated into the organic material by the action of the high energy at least with x-rays — I'm not sure what your UV source would give you.

HUBBARD: The principles of detecting life in our experiment is to expose the soil to the gases in the presence of a simulation of the Martian solar spectrum, sample the soil, pyrolyze, discard the pyrolysis CO_2 fraction (all the highly volatile materials) and look for the label in the organic fragments in the pyrolysate.

We may have exchange or addition of CO_2 into any one of a number of compounds, but in our method of analysis that CO_2 is discarded in the pyrolysis CO_2 fraction; we detect the label in the pyrolysate.

LEMMON: As methylene groups, and not carboxyl groups?

*See addendum, page 30.

HUBBARD: Not as carboxyl groups.

BADA: Is there any decomposition of the organic material in the soil? I think you said you get five per cent after 140 hours.

HUBBARD: Excuse me! I misunderstood your question: we haven't measured it.

MILLER: Have you looked at the compounds in the gas phase?

HUBBARD: Just very briefly. We passed the gas phase over our fire-brick column used in our pyrolysis experiment. It would have to have been something rather than polar to have been retained on the GC column at 120°C.

MILLER: For example, do you produce any methane?

HUBBARD: Honestly, we don't know yet. Jim Hardy is doing additional experiments, looking critically at the gas phase.

MILLER: What is the scale of synthesis for C^{14} incorporation? You have 10 mg of soil? How many milligrams of organic compounds would be synthesized in this?

HUBBARD: We're in nanogram quantities in most experiments. Twenty nanomoles is a good yield on the Vycor. On 30 mg of Vycor, less than 20 nanomoles; on 500 mg I believe we got 20 nanomoles, but I have forgotten the precise figures. Cyril [Ponnamperuma] is beginning some experiments where he is going to try to scale up the products.

MILLER: If you scale up the gas phase from 10 cc to 50, do you get the same percentage in synthesis?

HUBBARD: Here I have to relate some old experiments to some new ones. The experiments with a larger vessel (50 cc) were irradiated with the unfiltered Xenon source, with soil covering the bottoms. We appear to get about the same rate of synthesis per unit gas volume, as in the 5 cc experiments.

ORGEL: Then you have increased the surface area that you are irradiating at the same time?

HUBBARD: Yes.

MILLER: I'm trying to understand if the synthesis is limited by the amount of CO or by the surface.

HUBBARD: In more recent experiments where we used filtered xenon UV ($\lambda > 2100$ Å) the organic yield is proportional to the amount of surface and not to the gas volume.

Because we're dealing with small yields of organic material, the only way we could identify the organics is by following their radioactivity. First we ran an aqueous extract of an irradiated soil or Vycor sample on a thin layer chromatograph and then exposed these chromatograms to x-ray film to form an autoradiograph. We see about four or five products. If we expose for longer periods we can detect a couple more. About 70 to 80 per cent of the total radioactivity is in one spot. This spot is only seen on a thin layer chromatogram developed in an alkaline solvent system. Our major spot has identical R_f values (migration in this chromatographic system) to acetaldehyde and formaldehyde, which are inseparable. The only other spot we have tentatively identified appears to be glycolic acid ($CH_2OHCOOH$).

USHER: Why do you claim it is volatilized in an acid system?

HUBBARD: It just doesn't stay on the TLC [thin layer chromatograph] plate.

USHER: That's probably how these anions can be made, but that's probably a very small percentage. Formaldehyde can be hydrated, but that would be true in acid too.

HUBBARD: You can take known ^{14}C-formaldehyde or known ^{14}C-acetaldehyde, and they are volatilized in the acid solvents.

LEMMON: These compounds are gaseous at room temperature, the products themselves, but in alkaline solution I think you form trimers. Metaldehyde is one of them.

MILLER: What's the solvent?

SAGAN: That's the question.

HUBBARD: For example, it could be chloroform and ammonium hydroxide, or methanol with water, saturated with ammonium carbonate.

MILLER: The basic solution is always ammonia?

HUBBARD: No. The aldehydes are retained on a chromatogram developed in lutidine solvent. I expected them to be volatilized.

WOLMAN: What other indication have you that it's an aldehyde?

HUBBARD: Jim Hardy took these materials, reduced them with lithium aluminum hydride, mixed in some carrier unlabeled alcohols and passed these over a gas chromatograph. The products he identified were ethanol, apparently from acetaldehyde, methanol, which apparently came from the formaldehyde, and 1, 2-ethanediol, which apparently came from glycolic acid. We can't eliminate the possibility that the glycolic might have been the aldehyde, glycolaldehyde. We tried to keep our work-up under anaerobic conditions, but the aldehyde could have easily been oxidized to glycolic acid. Also, some of the methanol that was detected after lithium ammonium hydride reduction could have been formic acid or methanol.

USHER: Because borohydrides are not reduced, you could use them instead and prove it's not an acid.

HUBBARD: Hardy also saw one more product after the reduction. He had a small amount of n-propanol, which could have come from propional-dehyde or propionic acid. So we did an experiment where known aldehydes were mixed with the C^{14} extract, and the 2,4-dinitrophenyl (DNP) hydrazones were formed. He then recrystallized the derivatives several times with each of the three carriers added; formaldehyde, acetaldehyde, and propionaldehyde. The 2, 4-DNP crystals maintain a constant radioactivity on recrystallization. This is usually considered evidence for the various aldehydes.

SAGAN: Just a parenthetical remark. If you are making hexamethyl-tetraamine— as I guess a lot of people are — we can make methanol by ultra-violet irradiation of hexamethylamine. Possibly there are other alcohols that can be made this way.

MILLER: There are specific colorimetric tests that can be done on a very small scale for formaldehyde and acetaldehyde. Chromotropic acid (dihydroxy naphtahaline disulfonic acid) is used for the formaldehyde, and p-hydroxy diphenyl for the acetaldehyde. These might be of some use.

HUBBARD: We have concluded that yields are too low to identify prod-ucts by physical means until we scale it up somehow. We have only used radio-activity for our identifications.

What happens when various components of the system are limited? First we limited the level of carbon monoxide where the CO_2, water vapor and the soil substratum were held essentially constant (Fig. 3). The lowest concentra-tion of carbon monoxide tested was a partial pressure of 0.2 millibars. [Illus-trating] Over this 275-fold range we see only about a 6-fold reduction in the amount of organic material formed when we reduce the concentration to 0.2 millibars of CO. As for the conversion of CO to CO_2, the decrease is about 2.5-fold.

HUBBARD: Between 94 and 99 per cent CO_2 in these experiments.

SAGAN: And the rest CO?

HUBBARD: No, water vapor is about 2.8 per cent and CO varied from 0.2 up to 55 millibars partial pressure.

SAGAN: How valid are these experiments for Mars? Both the CO and water abundances are far in excess of the martian values.

HUBBARD: Yes. If you assume that the CO is 0.2 per cent of the 6-millibar atmosphere that would be 0.01 millibars of CO. Our lowest value is still twenty times this level.

SAGAN: You are even more generous with the water.

HUBBARD: True. Recently we ran some experiments with CO con-centrations lower than in this figure; 0.01 millibars of CO — in a system with a total pressure of 10 millibars, the balance being made up by CO_2 and water vapor in this same wet condition. We still saw organic synthesis.

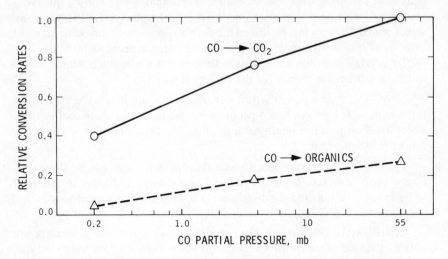

AFTER 17 HOURS XENON IRRADIATION OF CHAMBERS CONTAINING 30 mg
ORGANIC SOIL, 2.8 PERCENT H_2O VAPOR, $^{12}CO_2$ AND INDICATED
^{14}CO LEVEL.

Figure 3. UV conversion of carbon monoxide and water into organic carbon with CO limiting.

SAGAN: That kind of mixing ratio is correct. It is in the martian range, but your amount of the water is larger than on Mars.

HUBBARD: Right, but we have reduced the water in still another experiment. It is encouraging (from a martian standpoint) that the yield doesn't fall off too sharply with that concentration change. We don't know exactly what the low water level is because of the possible contribution of adsorbed water.

SAGAN: Do you have any quantitative figures: do you know what the water vapor mixing ratio is in the lower water experiments?

HUBBARD: Not really. We switch over to nitrogen as the carrier gas for that experiment and work on both sides of the system with liquid nitrogen traps. We do not intentionally equilibrate the Vycor with water vapor, but briefly pass it through the lab air. We then pump on the Vycor in the quartz tube several times. We see about a 10-fold reduction in rate of CO to CO_2 conversion, and also in the CO to organic conversion when dry nitrogen is the carrier gas.

FERRIS: What do you think is absorbing the light? Is it the water vapor which is far in excess of the carbon monoxide that is absorbing the light?

I have two questions: What's absorbing the light (1); and (2) with no CO at all does anything happen? Is the CO essential for organic synthesis?

HUBBARD: Yes, with no CO nothing happens. Water is probably absorbing the light. We probably have a photolysis of water with the UV below 2000 Å and the OH causes the oxidation of CO to CO_2. There probably are two reactions going on; one a reduction to the organic from the reducing potential generated from the photolysis of water, and the oxidative reaction.

SAGAN: Please let me ask about your diagram (Fig. 1B). Water doesn't start absorbing until about 1800 or 1850 Å, but on your solar curve, you have reached zero by there. I'm worried about the peak in your xenon source at 1800 Angstroms, which does not correspond to the peak in your solar system.

HUBBARD: Later I will show some data in which we filtered that UV from the spectrum.

FERRIS: But the point is the solar light doesn't end there. There is plenty of light intensity below 1800 Å.

SAGAN: But at some point you get to such low fluxes that there is no detectable photochemistry. The question is: where is that?

You see, in reality that's an exponential dip on the shortwave length side of the Planck distribution, so energy is lost very rapidly.

MILLER: But the photoefficiency there is much greater.

BADA: Have you exhausted the CO and followed the photochemical decomposition?

HUBBARD: Yes. I'll come to that. Figure 4 shows the absorption coefficient for water vapor at 2000 Angstroms to be about 10^{-3} as Thompson, Harteck and Reeves (1963) have shown.

SAGAN: Are you saying there is continuum absorption by water?

HUBBARD: That's how it's expressed in their plots. CO_2 cuts off a little shorter. I don't have the complete absorption spectrum of CO but the primary absorption is at about 1400 Å. With the xenon source we're providing UV that is absorbed in the Cameron bands.

FERRIS: But the Cameron bands of CO are not absorbing very much light. They are spikes, and they are not very high intensity versus your water. I don't know. It just seems to me that the concentration of water would be critical in determining your results.

HUBBARD: The last Cameron band at 2060 Å has an absorption coefficient of about 10^{-2}.

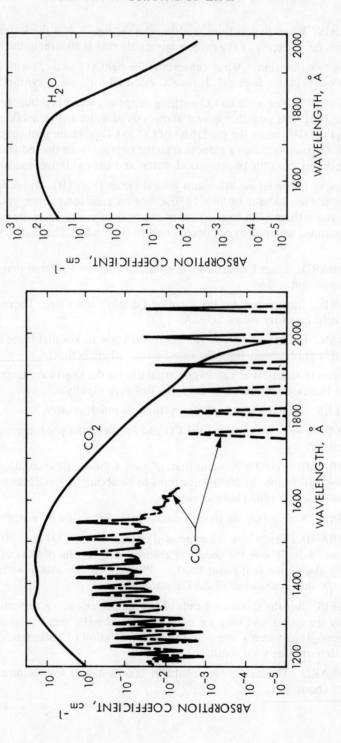

Figure 4. UV absorption for carbon monoxide and dioxide (left) and water (right).

SOFFEN: Before you finish this, is there any qualitative change in the organics?

HUBBARD: No, we have only observed quantitative changes. Our first identifications were made on extracts from irradiated soil and thereafter the identifications were made by comparisons on thin layer chromatograms. So in all conditions tested we see the aldehyde spot on the thin layer chromatogram and also the glycolic acid spot. With Vycor as the substratum, the proportion of glycolic acid always seems lower than in extracts from the soil substrate. With nitrogen as the diluent gas, we see the same products in about the same yield as we would have seen had CO_2 been the diluent gas. In the dry nitrogen experiments, the limited water experiment, we still saw the same products at the reduced yields.

SOFFEN: Did you use just free hydrogen?

HUBBARD: No. We have introduced ammonia.

MARGULIS: Aren't you putting in some organics and enzyme catalysts when you start with rich soil?

HUBBARD: Substantial amounts of organics and limited amounts of active enzymes.

MARGULIS: Vycor is sterile silicate material, and the soil must be full of all sorts of microbes that you assume you kill. It seems confusing to me.

HUBBARD: Yes.

MARGULIS: But that's an assumption. Have you tried to inoculate them?

HUBBARD: You don't completely sterilize soil by heating it overnight at 175°. It's hard to detect the few microorganisms that survive.

MARGULIS: But all it takes is one to grow and form organics.

HUBBARD: But whether it's Vycor, soil, the walls of the tube, or gold-plated silver wool we still detect the same products.

ORGEL: It's unlikely to have anything to do with organisms: presumably it doesn't work when the UV light is turned off.

MILLER: What are the products when ammonia is added?

HUBBARD: We have one new unidentified product which comes and goes. I haven't figured out the optimum conditions for its production.

MILLER: Is it glycine?

HUBBARD: No. About all we know is that it is not volatilized in either acid or alkali.

SAGAN: Leslie, may I propose a hypothesis which matches your synthesis hypothesis when the light goes off?

Let's suppose that the microorganisms that Jerry [Hubbard] can't kill are busy incorporating his labeled carbon gases. When the UV light is on these cells are being destroyed and their contents spewed out. When the UV light goes off, the cells are not being photolyzed; they retain their contents inside their cell walls, and therefore no increase in the amount of organics is observed. What's wrong with that?

ORGEL: But do cells break when you shine UV light on them?

MILLER: They are killed, but does the membrane rupture?

SAGAN: All bonds are broken at 1800 Å.

ORGEL: It's not clear to me that when cells are killed everything escapes.

SAGAN: I don't say everything; but certainly some things: volatile gases —

BADA: Is it realistic for enzymes to survive at 165°C?

LEMMON: Not as enzymes.

HUBBARD: Not at 720°C on Vycor.

ORGEL: We must distinguish what's likely to be happening and what's just conceivable. Carl, what you are saying is just conceivable, but on the basis of what's likely to happen, I don't think its likely.

LEMMON: I'm still bothered by this whole discussion, and I hope some other person here is also.

Supposedly we are discussing new prebiotic synthesis. What is a new prebiotic synthesis?

HUBBARD: The new aspect of this is that we are forming organic compounds.

LEMMON: But formaldehyde and acetaldehyde were observed twenty years ago by irradiating CO and CO_2.

SAGAN: The new aspect is the use of long wavelength ultraviolet light.

HUBBARD: The synthesis occurs at wavelengths beyond which the gases in our system absorb, at wavelengths which are presently reaching the martian surface, and which could have been reaching the surface of the prebiotic earth.

LEMMON: But is it fair to say that you have not synthesized some new compound that has not been observed before under prebiotic conditions?

HUBBARD: No.

MILLER: But it is important to know if large quantities of previously synthesized compounds can be synthesized by another process.

LEMMON: By the presence of these known organic compounds, or even cells in the soil.

MARGULIS: These experiments start with a huge quantity of unknown organic matter. I don't see how you can get anything out of it.

MILLER: But he got good yields of organic compounds on Vycor.

MARGULIS: Then let's look at the Vycor, and not the soil.

HUBBARD: I wish I had complete data on the Vycor to simplifiy things. You can see how the problem evolved; first seeing the phenomenon and then trying to make it go away to save the life detection experiments!

FERRIS: What is the physical organization of Vycor?

MILLER: Vycor is made by taking Pyrex, leaching the boron (borax) out of it. It is then quite spongy; the next step is to fuse it. The result is not pure silica, but something containing about 98 per cent silica.

MARGULIS: Is it small particles with great surface?

MILLER: No, during the fusion step, the spongy material contracts so there are no holes.

SOFFEN: No, that's historical. They stumbled into it.

HUBBARD: No. It has some importance in being opaque.

SAGAN: Under those circumstances I agree with Leslie Orgel's comments to me; you don't have to assume microorganisms.

HUBBARD: Table 1 shows the percent conversion of CO to CO_2 with just the quartz tube, no Vycor in it, with 30, 150, and 500 mg of Vycor. The atmosphere was 0.06 per cent CO plus diluent gas, CO_2, and water vapor. Without Vycor, after 17 hours irradiation about 90 per cent of the CO has been oxidized to CO_2. An aqueous extract of the walls of the vessel indicated that about 1.8 per cent of the CO has been converted to organics.

SCHOPF: Are the organics labeled?

HUBBARD: Yes, the organics were produced from the ^{14}CO. If we extend the radiation time to 69 hours, we get more conversion of CO to CO_2, and see a decrease in the yield of organics extractable from the walls of the quartz vessel (Table 1). Once the CO has been depleted photo-destruction proceeds. After 69 hours only 0.4 per cent organics remain in the vessel containing no substratum.

With 30 mg of Vycor and 69 hours irradiation we see about 50 per cent conversion of CO to CO_2 and 3.15 per cent conversion to organics. With more Vycor – 150 mg – about 20 per cent conversion of CO to CO_2, but roughly the same yield of organics in 17 hours.

TABLE 1. EFFECTS OF VARYING LEVELS OF SUBSTRATUM ON THE UV-CONVERSION OF ^{14}CO TO $^{14}CO_2$ AND ^{14}C-ORGANICS

Vycor* (mg)	Xenon UV (hr)	Radioactivity** in		
		^{14}CO	$^{14}CO_2$	^{14}C-Organics
30	None	60.5	1.5	0.0005
None	17	3.8	45.0	0.91
None	69	1.4	47.8	0.21
30	17	25.0	27.8	1.72
150	17	40.1	10.8	1.28
150	69	19.6	33.0	3.42
500	163	8.1	32.3	5.33

*5.5 cc quartz chambers contained the indicated quantity of Vycor particles and a gaseous mixture of 0.06% ^{14}CO, 2.8% H_2O vapor and 97% $^{12}CO_2$.

**Expressed as counts per minute $\times 10^{-5}$.

With 500 mg of substratum irradiated for 140 hours there is a well over 10 per cent conversion of CO to organics.

ORGEL: Is what's left over still CO? I suppose you get 10 per cent converted to CO_2, and 3 per cent into something else, so it always adds up?

HUBBARD: Yes. As I said, there could be traces of other gases present.

ORGEL: But more or less it adds up.

HUBBARD: Yes.

SOFFEN: What is the transmissibility of the Vycor to the irradiation?

HUBBARD: We haven't measured it.

SOFFEN: It may be cutting off?

HUBBARD: Yes.

SAGAN: At these wavelengths you would use quartz — because Vycor does not transmit light.

SOFFEN: But when you increase amounts of Vycor, do you maintain the same cross-section? Do you fill the vessel more, or have a larger area?

HUBBARD: When we add greater amounts of Vycor to the vessel we expose more surface area, but the increase is not proportional to the weight of substratum. Slowing down the conversion of CO or CO_2 doesn't speed up the rate of organic formation, but it keeps CO in our system longer, to permit us to build up a better yield of organics on extended irradiation.

SAGAN: Jerry, please explain the geometry — is it an end window? What contains the gas?

HUBBARD: A quartz tube, 5.5 cc volume, with the quantity of Vycor spread over the bottom.

SAGAN: So short wavelength UV gets in and is strongly absorbed by the Vycor; because Vycor isn't used at short wavelengths for transmission — we know it's a good absorber.

HUBBARD: Yes.

LEMMON: What happens in the same conditions without Vycor?

HUBBARD: This was shown (Table 1). After 17 hours there is 1.8 per cent conversion to organics. After 69 hours it falls down fivefold, since we have depleted the CO and are destroying the previously synthesized organics.

HULETT: Are you irradiating from the side? from the front?

HUBBARD: All of the experiments described thus far have been irradiated from above.

HULETT: So the UV goes through the gas and then hits the Vycor?

HUBBARD: Yes.

FERRIS: Have you used powdered quartz? In other words, this is a question that has been raised here. Is the Vycor absorbing the energy and somehow activating the gases?

HUBBARD: I'll get out of turn here. The tube can be placed vertically with the Vycor in the bottom, and shielded from the direct UV coming from the front. With the unfiltered xenon lamp there is an appreciable yield of organics formed on the shielded Vycor.

SAGAN: So, then, the absorber is not the Vycor.

HUBBARD: With the Vycor substratum shielded from the direct UV there is a little better yield than that which receives the direct irradiation.

BARGHOORN: Why do you go from soil to Vycor, instead of to montmorillonite, or some other naturally occurring —

HUBBARD: We have just started some experiments with a variety of substances, but this was a nice, clean system which worked.

SAGAN: I can see the urge for a nice, clean system.

HUBBARD: For calculation purposes we used 75 meter atmospheres of CO_2 as the column the solar UV would pass through to reach the surface of Mars (Fig. 1A). We calculated that the solar UV down to about 2000 or 1950 Å, should reach the surface. To simulate this spectrum we filter the xenon UV through five centimeters of liquid water or methanol. I'm defining as efficient attenuation the amount which gives an O.D. of 2. Thus, the water filter should effectively remove the UV below 1900 Å and the methanol filter that below 2100 Å (Fig. 1A). We have also used pyrex to filter out the UV below 2800 Å.

Table 2 gives the conversions of CO to CO_2 and to organics after 17 hours of irradiation. With the 5 cm quartz cell containing air, we see almost 50 per cent conversion of CO to CO_2 and 1.6 per cent conversion to organics.

TABLE 2. ORGANIC SYNTHESIS WITH FILTERED XENON UV

Filter	% Conversion of ^{14}CO to	
	$^{14}CO_2$	^{14}C-Organics
None	47.7	1.6
Water	1.9	0.36
Methanol	<0.4	0.1
Pyrex	<0.1	0.02

5.5 cc quartz chambers containing 30 mg of Vycor and a mixture of 0.06% ^{14}CO, 2.8% H_2O vapor, and 97% $^{12}CO_2$ were irradiated 17 hours.

If we use the water filter, we reduce the conversion of CO to organics down to about 0.4 per cent and cause an even greater decrease in the amount of CO that's converted to CO_2.

If we cut off more of the UV with methanol, we still get about 0.1 per cent conversion to organics. This level of CO_2 is hard to quantitate: it is probably less than 0.4 per cent. These low yields are not comfortable to work with. If we add more Vycor and extend the irradiation time, we can obtain organic yields of over one per cent.

SOFFEN: In how long a time?

HUBBARD: 3 days.

SOFFEN: Is it still going up?

HUBBARD: Yes. It can be run a long time, because the CO is not rapidly oxidized to CO_2. As I previously mentioned the products that are formed using the filtered UV appear to be the same as are formed from the full xenon spectrum.

NAGYVARY: What sort of absorption filter is the methanol?

HUBBARD: In our system 5 cm methanol removes the UV below 2100 Å (Fig. 1).

SAGAN: This implies that with the methanol filter there is only radiation longwards of 2100 Angstroms entering your system?

HUBBARD: Yes.

SAGAN: And what is absorbing at wavelengths longwards of 2100 Å?

HUBBARD: Surface-gas complexes must be absorbing at the longer wavelengths.

SAGAN: CO_2 doesn't.

HUBBARD: No.

SAGAN: CO doesn't either.

HUBBARD: Terinin (1959) has shown that water adsorbed to silicate can be photolyzed at 2300 Angstroms.

HUBBARD: It must be a surface-water complex or maybe a surface-CO complex. We don't really know.

MILLER: If there were some iron on the surface, it could absorb and transfer —

SAGAN: Vycor is still absorbing somewhere at 2100 Angstroms.

ORGEL: I think, Carl, the suggestion is that when you absorb water on the surface, the natural UV spectrum of the water is modified. The suggestion (I don't know whether it's right or not) is that it's not due to the Vycor, but a genuine effect on the water. There is evidence in the literature showing that the spectrum of molecules are modified when they are associated with a surface.

SAGAN: Is this originally the work of Terenin?

ORGEL: Terenin was the first to do it, but others have too. I think zinc oxide has a particularly active surface.

HUBBARD: There are many other examples. If I may be permitted to go back to experiments with organic soil, with UV ($\lambda > 2100$ Å) the reaction doesn't work as well with soil and with Vycor. We're apparently not providing the appropriate surface-water or surface-CO complex by using the organic soil substratum. We have tried one more material which is similar to a basalt. It is latite, a fine-grained material or volcanic origin, which is less basic than a basalt.

SOFFEN: Is that on a per weight or a per area basis?

HUBBARD: Per weight. It is between 1 to 5 per cent as efficient as Vycor under these conditions.

SOFFEN: Yes, but clearly you are talking about a surface, not a bulk phenomenon.

HUBBARD: Yes.

SOFFEN: Should you not compare everything per square centimeter?

HUBBARD; Yes, and we should compare several other parameters. Clearly we need to consider several models of the martian surface thing before concluding this reaction has significance on Mars. I have mentioned three possible limiting factors, when considering Mars. Is there adequate water vapor, CO, and appropriate surface material present?

SOFFEN: Suppose you permit the experiments to run a long time. Is everything photolyzed? Do you recover inorganics?

HUBBARD: I have never done it.

SOFFEN: It seems relevant to Mars. You ought to look for equilibrium.

HUBBARD: You would keep photolyzing organics to recycle CO. What is the equilibrium on Mars, when CO is continuously produced by photolysis of CO_2 in the upper atmosphere? I don't know the diffusion rates to the lower atmosphere, but it seems as if CO would always be present even with our reactions occurring on the surface.

SOFFEN: You need a continuous source of CO — long term experiments to find out how much organic ultimately could be generated without photolysis.

HUBBARD: That will be done.

SOFFEN: What worries me is that when you get to Mars you will find something that had nothing to do with Mars.

HUBBARD: Even with Pyrex filters which absorb down to about 2800 Å, there is a detectable level of organics on the Vycor.

ORGEL: If the Vycor were absorbing, I could understand this, but I can't believe that either water or CO, even adsorbed on the most suitable matrix, would absorb light that far out.

HUBBARD: Yes, it's asking a lot. That's why I was encouraged when the latite did not work very well with the long UV.

ORGEL: Can you tell if a great big block of your material on a very thick long path is absorbing any — ?

HUBBARD: Perhaps we might get some from the manufacturer.

LEMMON: Could the residual synthesis just be a matter of the radiation of the carbon[14] beta particles? The radiation dose seems pretty small, what is it? How many microcuries of CO have you?

HUBBARD: But we compare experiments: the same sample in the same conditions in the dark —

LEMMON: Without the light don't you get anything?

HUBBARD: Nothing. We have also used a low pressure mercury lamp; its primary emission is a strong line at 2537 Å, another at 1849 Å, and a very small line at 1942 Å (Fig. 1). The next line of any significance is out at 2135 Å. With this source, the water or the methanol filter is used to chop off the 1849 line, and with the methanol system you get 36 per cent, less than 0.6 per cent organics (still 0.72 per cent conversion) about 70 per cent of the maximum with the water filter. (See Table 3)

TABLE 3. ORGANIC SYNTHESIS WITH FILTERED, LOW PRESSURE MERCURY UV.

Filter	% Conversion of ^{14}CO to	
	$^{14}CO_2$	^{14}C-Organics
None	36.1	1.03
Water	<0.6	0.72
Methanol	<0.4	0.48
Pyrex	<0.1	<0.002

5.5 cc quartz chambers containing 30 mg of Vycor and a mixture of 0.06% ^{14}CO, 2.8% H_2O vapor, and 97% $^{12}CO_2$ were irradiated for 17 hours.

Table 3 shows another source of energy longer than 2100 Å giving a result. We have also used an iodine lamp with a methanol filter that has primary emission lines at 2060 Å, but some others too. We still detect organic formation at energies between 2100 Å and 3000 Å.

MILLER: Let us for a moment consider the primitive earth rather than Mars. Do you believe this sort of UV synthesis was very important relative to other proposed syntheses of the same compounds? Considering all the variables — atmospheric components that might absorb the light, the amounts of exposed surfaces and so on — would your syntheses have been more effective than the old-fashioned electric discharge?

SAGAN: What are the quantum yields?

HUBBARD: We have no estimates.

SAGAN: But the manufacturers' specifications for these lamps are usually correct to within two orders of magnitude!

MILLER: I thought you said it was 37 watts per square centimeter.

HUBBARD: I said the total irradiation from this source is 32 milliwatts compared to 42 for the martian surface at zenith: IR, visible and UV.

If we take about 0.5 milliwatts per square centimeter times the 100 Å spread total active energy will fall below 3000.

SAGAN: A quantum yield calculation should be possible from that.

USHER: But you must know how much energy is absorbed by what.

SAGAN: Not just to get an effective quantum yield.

USHER: Yes.

SIEVER: Have you used metal sulfides and oxides, which might be more instructive.

HUBBARD: No, we have only used one soil sample, one volcanic sample, and glass.

SAGAN: Despite the fact that David [Usher] says it's useless, let me try to calculate the quantum yield. How many molecules of organics do you make?

HUBBARD: How many molecules? In a 17-hour experiment we had 0.06 per cent CO in a 5.5 cc system — and we get about 3 per cent conversion in a good experiment. The organics derive their carbon from CO. I think the CH_2O product is useful to calculate that.

SAGAN: How many grams of formaldehyde? I'll get the total number of photons if you give me the total number of molecules. A quarter of a micromole?

HUBBARD: About 10 nmoles times the molecular weight of 30.

LEMMON: You just want molecules. With 10^{-8} moles you have about 10^{15} molecules.

SAGAN: 10^{15} molecules from 10^{20} photons, so it's a quantum yield of 10^{-5}, which is highish. By highish, I mean it's significant.

MILLER: It's high compared to Groth and von Weyssenhoff (1957).

SAGAN: It's a little high compared to Groth and von Weyssenhoff, who used mercury sensitization, but it's low compared to our H_2S and formaldehyde sensitizations for the production of amino acids (Khare and Sagan, 1971). Von Weyssenhoff's and our uv photoproductions are directed at amino acids, and you are reporting only aldehydes. Still, it's a very respectable quantum yield.

USHER: It's a minimum, though.

SAGAN: Yes, that's why I say it's meaningful. If it were 10^{-3}, it would be even more interesting.

SOFFEN: May I ask you about the xenon lamp? Do you exclude oxygen? Does it generate ozone?

HUBBARD: There is about five inches of column of air that the lamp goes through, but it's being swept.

SOFFEN: So there is no ozone filter. Has the fan ever gone off, and an ozone filter been built up?

HUBBARD: No. We have taken the quartz cell (usually with the 5 cm of water or methanol) and used it with air and kept it closed.

SOFFEN: That you have a good ozone filter is the more powerful argument.

HUBBARD: Not a very good one — we made some calculations.

SAGAN: Maybe I can put some calculations on stability on Mars on the board.

Aldehydes on Mars are not a new issue. It's an old, agonizing problem because of the so-called Sinton bands. These were infrared absorption features in the 3.5 micron region of the Mars reflection spectrum reported by William Sinton more than ten years ago. They were attributed to a CH stretch transition; the only molecule that seemed to match, particularly the long wavelength band, was acetaldehyde. At that time I was concerned about the stability of acetaldehyde on Mars, and I think I'll be able to reproduce the calculations here.

I have a set of curves of the acetaldehyde absorption spectrum; the main point is that its peak absorption is at quite long wave length around 2900 Å. The absorption cross-section is several times 10^{-20} square centimeters.

First, let's ask if there is self-protection of acetaldehyde on Mars. That is, is there enough so that it protects itself from photodissociation, so that acetaldehyde at the lower altitude can survive?

From ultraviolet spectra of Mars taken from the orbiting astronomical observatory [OAO] we do happen to have some upper limits on the acetaldehyde and the formaldehyde abundances (Owen and Sagan, 1972). This is unpublished but I'm sure it will be in print by the time this book is in print. Anything will!

ORGEL: An optimist!

SAGAN: The OAO results give upper limits on the abundance of acetaldehyde of about 1 mm atmosphere. A centimeter atmosphere is converted into molecules per square centimeter by multiplying by Loschmidt's number, 3×10^{19} cm^{-3}. However, I have converted it for you into grams per square centimeter. It's 10^{-4} grams per square centimeter which implies about 10^{18} molecules per square centimeter.

LEMMON: Does this say anything about acetaldehyde that can be absorbed in interstices of clay?

SAGAN: No. I'm only discussing the gas phase now. Instead of 10^{18} molecules per square centimeter, therefore, the optical depth — or I believe you say O.D. —

LEMMON: Optical density.

SAGAN: A dimensionless unit, right?

LEMMON: Yes.

SAGAN: Apparently we are talking the same language.

So you multiply the number of molecules per square centimeter by the number of square centimeters per molecule and you will get the optical depth. Okay? I call the optical depth tau. That optical depth is less than three times 10^{-2}. In other words, if there is this much acetaldehyde in the martian atmosphere, the photodissociating photons get to the very surface. The conclusion is even stronger for formaldehyde. What is the first order lifetime of any molecule in seconds? It is one over the flux in photons per square centimeter per second times the cross-section in square centimeters. At wavelengths around 3000 Å the photon flux is several per cent of the total photon flux at Mars. The solar constant gives a total photon flux about 10^6 ergs per square centimeter. For Mars in the ultraviolet it's several times 10^4 ergs per square centimeter per second.

At these wavelengths a typical photon has an energy of a few electron volts, several times 10^{-12} ergs. From that we derive a flux of 5×10^{15} photons per square centimeter per second. Multiplying by 10^{-20} square centimeters, we come out with the life time: it is about 10^4 seconds — a few hours. Therefore any acetaldehyde or formaldehyde molecule in the martian atmosphere is photolyzed by solar ultraviolet radiation in a few hours — less than a martian day.

LEMMON: Carl, (as the younger generation says) what's the relevance of all this? All the acetaldehyde may be adsorbed on the soil, very nicely protected.

HUBBARD: Intuitively (we haven't done calculations), we think any accumulation is going to be dependent on leaching of materials into the soil, or some dust storm covering them up, or something. And we have an added complication: Is there enough ozone?

SAGAN: There is certainly not enough ozone to protect aldehydes. Ozone is surely being made by CO_2 photodissociation — oxygen atom, oxygen molecule combination, but because the pressure is so low on Mars, that recombination occurs at the surface. The peak ozone abundance on Mars is at the surface.

This in fact is an old explanation of the red color of Mars: the ozone oxidation of more reduced iron minerals to iron oxides.

MILLER: Did Mariner flights yield information on the lower limit?

SAGAN: Yes, and so did OAO (Orbiting Astronomical Observatory). Scanning the spectrum we see a little dip at 2600 Å. Barth's experiments give a feature consistent with ozone. However, it's also consistent with the monomer of carbon suboxide — very like ozone.

MILLER: Was any ozone detected in the infrared?

SAGAN: No, but the electronic transitions are many orders of magnitude stronger.

ORGEL: Knowing the quantity of ozone can the amount of oxygen be estimated?

SAGAN: Yes, indeed.

ORGEL: Is it consistent?

SAGAN: Yes, with the upper limits. It's also consistent with the theory of CO_2 photodissociation, which produces a certain amount of oxygen that eventually makes a certain amount of ozone.

ORGEL: This suggests that the experiments be repeated with CO and oxygen. Not chemically, of course, but in photochemically equivalent amounts.

HUBBARD: We did an experiment with half the O_2 than we had CO; the yields were the same.

SAGAN: O_2?

HUBBARD: And CO_2 and CO.

SAGAN: Yes, but not ozone.

MARGULIS: What are these upper limits of ozone and O_2?

SAGAN: O_2? There's a claimed identification and a claimed refutation, (Belton and Hunten, 1968; Margolis, Schorn and Young, 1971), so it's hard to know. The figure there is some tens of centimeter atmospheres claimed for oxygen, and the ozone upper limit is about 10^{-2} cm-atm. I'm not absolutely sure.

MILLER: What's that in millibars or mixing ratio?

SAGAN: The total martian atmosphere is like a hundred meter atmospheres, so there's 10^4 cm-atm — implying 10^{-6} mixing ratio for ozone and something like 10^{-3} for oxygen.

BADA: What half life do you calculate for formaldehyde?

SAGAN: The same.

BADA: What half-life do you calculate for formaldehyde for interstellar space?

SAGAN: The formaldehyde lifetime in interstellar space is about thirty years.

BADA: This implies that to observe it you would have to have almost a similar synthesis rate? There you believe formaldehyde is synthesized at that rate?

SAGAN: Absolutely. I believe virtually all the molecules we see in the interstellar medium are generated in situ from the grains and haven't migrated from some other place. This is a whole other issue.

LEMMON: Is it not perfectly all right to think of there being lots of acetaldehyde in the martian soil?

SAGAN: Even if it is in the soil. The photons get to the soil.

LEMMON: Not necessarily.

SAGAN: Do you want the molecules to hide under opaque rocks?

LEMMON: Yes. Protection wouldn't take much soil, would it? A centimeter would be plenty.

SAGAN: Much less. 100-micron range particles suffice.

LEMMON: So acetaldehyde there may be involved in chemical evolutionary processes. Your calculation shouldn't convince us the acetaldehyde is all destroyed.

SAGAN: All the acetaldehyde can be placed where it can be seen optically, yet UV photons can't get to it — but this is not a testable hypothesis.

LEMMON: Not yet.

FERRIS: If you adsorb acetaldehyde — on the surface (is this what you are saying?) you approach an equilibrium of acetaldehyde on and off the surface. With that strong photon flux a bit may be preserved on the surface, but if there is an equilibrium eventually it will shift away —

LEMMON: It's not an equilibrium process. It's getting adsorbed in the soil and protected from ultraviolet light; in the soil maybe radioactivity will do something else to it. There are billions of years to work with.

SAGAN: I agree with Ferris' remark. It's comparable to evaporation into a vacuum. Some solid with high vapor pressure can be maintained if there is a back reaction; but if every time a molecule comes off it is grabbed and taken away — then soon the entire amount evaporates.

FERRIS: The reaction in the soil must be faster, it seems to me, than the photolysis rate.

LEMMON: Higher molecular weight material in the soil may slowly accumulate from the acetaldehyde adsorbed there.

SOFFEN: It also freezes out every night potentially increasing local concentrations.

SAGAN: So what? It vaporizes the next morning.

FERRIS: But there are still some hours in which any number of things can occur.

SAGAN: Now that Jerry (Hubbard) has calculated his production rates, let me do one last thing.

In steady state the production rate equals the destruction rate. This implies we can calculate the abundance, as production rate, times the time. 10^{15} molecules cm^{-2} were made in 10^5 seconds, using the uv solar flux at Mars; this is 10^{10} molecules cm^{-2} a second.

Multiplying 10^{10} molecules cm^{-2} a second by 10^4 seconds, gives me 10^{14} molecules per square centimeter column. That is, under your assumptions, on the production rate, we can set an upper limit on the abundance four orders of magnitude below the observational upper limit. There's no contradiction. Because we don't see 10^{18} per square centimeter doesn't mean you must be wrong. There may be a tiny amount: the equilibrium abundance between a rapid production rate and a rapid destruction rate. Under no circumstances will there be much atmospheric acetaldehyde.

ORGEL: If some is hidden away in the soil, and some in the atmosphere, then there may be another ratio for that in the soil to that in the atmosphere.

SAGAN: Certainly, and the total still may be under the observational limits.

HUBBARD: Did I mention something pertinent to what Dr. Lemmon said? That is, the worst limitation put on the production rate was the reduced water level. When we used organic soil as the substratum, we accumulated a nice yield of ^{14}C organics at about less than one-tenth the rate that they were produced in a water-sufficient system.

BARGHOORN: Would you please tell us what you mean by organics? Just what do you mean?

HUBBARD: I mean conversion of CO and water to aldehydes, glycolic acid and the unidentified products.

BARGHOORN: Are they all small molecules?

HUBBARD: We don't know anything about the unknowns. It has been shown that acetaldehyde can be polymerized with this UV source.

YOUNG: All kinds of sugars can be made – good, healthy sugars.

MILLER: Even though the discussion is lively, I'm sure that many of us are tired. Let us stop now for the evening; we will continue tomorrow.

ADDENDUM

Recent experiments have shown that formic acid is the major product of the surface-dependent reaction of ^{14}CO and H_2O. We erroneously assumed that formaldehyde was the major product based on the similarity of behavior of the unknown and authentic ^{14}C-formaldehyde. However, the commercial ^{14}C-formaldehyde was found to be heavily contaminated with ^{14}C-formic acid. ^{14}C-Formaldehyde has been detected as a minor product ($\leq 1.5\%$ of the total ^{14}C-organics) by preparation of 2,4-dinitrophenyl hydrazone and dimedon derivatives in the presence of excess ^{12}C-formaldehyde.

The major product of the surface-dependent reaction of ^{14}CO and NH_3 have been tentatively identified as ^{14}C-urea. The identification is based on thin layer chromatography-radioautography and digestion with urease.

J.S. Hubbard, October 1972.

ORGEL: As I am sure you all know, one of the plagues of prebiotic chemistry is the identification of very small amounts of material and problems of contamination. John Oró is an expert on this, so I'm just going to ask John to begin.

ORÓ: Am I first?

ORGEL: Have I made a mistake? I understood that I was to be Chairman and you were to work, John!

ORÓ: All right. This is certainly one of the most important problems in the study of the synthesis of biochemical compounds. All the data obtained and all the theories built on these data depend on the accuracy of identification of the organic compounds synthesized in experiments carried out under possible primitive earth conditions.

There are five particular areas where the problem of identification and contamination apply: (1) products synthesized in prebiological simulation experiments; (2) analysis of ancient sediments in attempts to determine the earliest evidence of life on earth; (3) analyses of carbon-containing meteorites, particular the so-called carbonaceous chrondrites; (4) lunar samples — rocks or lunar soil — where the analysis is to determine the organic as well as the organogenic elements and compounds, and their possible relationship to the problem of chemical evolution; and, (5) of course, other planetary surfaces, specifically the martian surface, which presumably will be analyzed in 1975.

What methods are available to answer the questions of correct identification and contamination? I am not going to discuss them in detail, but only to mention them briefly. There are actually several methods available, and particular problems may be solved better by one method than another.

I have listed about ten methods here. Basically, they fall into two categories: chemical and physical. (1) *physical constants,* such as melting point, molecular weight, optical rotation; (2) *optical measurements,* such as infrared, visible, ultraviolet, absorption or emission methods — including fluorescence and phosphorescence; (3) *field interaction spectra,* methods such as NMR, EPR, ORD; (4) *diffraction and related methods,* such as x-ray diffraction, electron diffraction, *et cetera,* including also, different imaging techniques, such as EM (electron microscopy), ESM (electron scanning microscopy), EMP (electromicroprobe analysis), IMP (ion microprobe analysis); (5) *fragmentation spectra,* such as mass spectrometry, which I have

31

listed here as EMS (electron mass spectra), and also, IMS (ion mass spectra), CIMS (chemical ionization mass spectra) and TMS (thermal mass spectra), such as Curie point and laser fragmentation mass spectra; (6) *relative separation methods.*

PONNAMPERUMA: Where would you put ESCA into this?

MARGULIS: What's that?

PONNAMPERUMA: Electron spectroscopy for chemical analysis. Isn't that your strong point?

ORÓ: It's one of the "et cetera" within group (4). It's certainly an important new development. I was listing here the methods being used that we have some familiarity with. It would take time to go into the ESCA.

Coming back to class (6), we have so many *relative separation methods* (based on partition, ion exchange, molecular sieving or other principles) that they can be classified into two major groups: methods in which the materials that are analyzed are gases or volatile compounds, for example, gas-liquid partition chromatography, gas-solid adsorption chromatography or methods that involve liquids or solids, dissovled in solvents. Of course, the solid may be a monomer, or a polymer. If it's a monomer, there are well-known simple methods of thin layer chromatography, paper chromatography, ion exchange chromatography, among others. If the solid is a polymer, we perhaps use more specialized separation methods: countercurrent distribution, ultrafiltration, ultracentrifugation, electrophoresis, etc., or ion exchange methods, in which derivatives of say cellulose, and other polymers act as filtering (molecular sieve) or ion exchange agents.

Class (7) comprises the so-called *nuclear and isotopic methods,* to differentiate them from the chemical methods. Here I'm just indicating techniques like isotope dilution, isotope ratios, radio tracer methods, natural radioactive analysis, and induced radioactive analysis.

Class (8) in my list comprises the *chemical reaction methods:* i.e., specific chemical reactions producing specific easily characterized products. For instance a compound such as adenine may be determined by the specific Gerlach-Doring test (Oró, 1960), or a functional group present in an organic compound, such as an aldehyde, a ketone or an amino group may be determined by a functional group specific test such as the ninhydrin reaction for amino acids.

Class (9) refers to the *biological methods.* The presence of a certain compound can be tested by growing an organism that specifically requires the compound as a nutritional factor. If the compound is indispensible for growth, a qualitative as well as a quantitative determination of the compound may be made by bioassay.

The last class (10), *integrated methods,* includes any method resulting from the integrated combination of those listed above. One of these methods which is particularly appropriate for consideration here is combined gas chromatography mass spectrometry (GC-MS) which embodies the advantages of one of the best separation methods and of one of the most rigorous molecular identification methods.

MILLER: Will you discuss specific cases?

ORÓ: Yes. Please let me come up with two general principles that, although highly oversimplified, may focus our approach to these problems.

In relation to the problem of contamination, I recognize the "relative biological uncertainty principle", i.e., there is a point of diminishing returns in the analysis of biological molecules when the sensitivity is increased beyond certain limits. Below a certain limit, in the number of molecules we are analyzing, we cannot be certain that the molecules measured come from the system being measured or the system performing the measurements. If we go down in the limit of detection, we reach a point where life (i.e. man) cannot detect life by looking at molecules produced by life, because, at very low levels, these molecules are practically everywhere on earth. Of course, these limits are not well defined, and vary with the nature of the compounds that we wish to measure. For a large number of methods the sensitivity beyond which a point of diminishing returns is reached is between 10^{-8} and 10^{-11} moles.

Roughly, for those accustomed to measurements in parts per million, this is the same as saying: between one part per million and one part per billion (I'm assuming a monomer here of a molecular weight of 100).

For other methods, including gas chromatography, these limits may be lower: from 10^{-11} to 10^{-14} moles. Only in exceptional cases of polymers can this limit be pushed to the absolute maximum sensitivity: the method that allows you to see one single molecule.

Now, you say, this is impossible. Yes and no.

MARGULIS: I think smell receptors are down to one molecule.

MILLER: Really? I thought it was about a thousand.

SCHOPF: If you don't smoke cigarettes or live in a smoggy atmosphere.

MARGULIS: It's better than the 10^{-11} moles.

ORÓ: A good electron microscope permits you to see a single molecule with your own eyes — a gene or nucleic acid of 1000 or 2000 nucleotides roughly equivalent to one or two microns in length.

The other absolute method, perhaps the most interesting I have seen, was developed in Professor Lederberg's lab. It measures by fluorescence a single enzyme molecule in a droplet where there are many molecules of substrate with organic groups attached that, upon a certain enzymatic cleavage, become fluorescent. The droplet then lights up and, by direct microscopic observation of a lit droplet surrounded by many unlit droplets, the presence of one single enzyme molecule within a droplet is inferred. Let's not discuss absolute sensitivity any further − if in our work there was only one molecule to be detected I would have a hard time finding it. I am just trying to point out that the sensitivity for identification varies greatly (between 10^{-8} and 10^{-23} M) depending on the methods used and that for each method and type of compound there is a limit of usable sensitivity.

MILLER: John, we need to be more specific −

ORGEL: We ought to be more concerned with specific problems of identification of prebiotic compounds.

ORÓ: I was coming to this.

SCHOPF: But please don't forget to mention your "relative biological uncertainty principle."

ORÓ: That applies to this problem of sensitivity, of the limitations of sensitivity in relation to the problem of identification. The problem I'm sure Stanley was more concerned about is: When are we certain that we have identified a compund? Okay?

Classical organic chemists use three different methods to identify a compound. If three different methods provide an unequivocal answer, then you have a high degree of certainty. Relative methods *per se* do not provide an identification and cannot be used as having demonstrated the presence of a compound. And now I will discuss a specific example.

The problem of amino acids in relation to prebiotic synthesis experiments and analysis of lunar samples has been widely discussed. Some work presented is based only on relative retention times. To me this is not a demonstration of the presence of these compounds. Perhaps this indicates they may be present but certainly does not prove it. Journals should not accept this kind of evidence *per se,* unless provided with other unequivocal identification. Analysis of amino acids in samples, whether terrestrial, extraterrestrial or the products of synthesis, can be done by ion exchange and gas chromatographic methods and by ninhydrin reaction, but none of these by itself is good enough. The ideal situation should be the isolation and melting point (mp) determination of these amino acids. An equally valid method of identification would be combined gas chromatography mass spectrometry of amino acid derivatives.

I don't want to discuss questionable identifications in the literature but we must be guided by this principle. If the evidence is of the relative retention type, it has to be confirmed by infrared spectroscopy, if available, or ultraviolet if it applies. Eventually it should be supported by melting point determinations. If quantities are too low for mp determinations one must use other methods which provide unequivocal characterization, such as mass spectrometry, which have a higher sensitivity, comparable to that of the best relative retention methods.

MILLER: But, John, this is still not specific enough — for some compounds more must be done than for others.

ORGEL: Stanley, please permit John to finish — Then I plan to discuss the various classes of compounds, and give people who have worked with them an opportunity to talk. Perhaps we can produce what I would call the "Belmont Manifesto" which would state ideal and minimal conditions for the identification of various classes of compounds! [Laughter]

SCHOPF: And make everybody sign it! I think we're about to be railroaded.

SAGAN: What about those who are not here, and don't sign it?

ORGEL: John has indicated the two sorts of problems: first, contamination. Do the different molecules observed come from where you think they came from, or from your fingers? Secondly, there are quite different problems of identification: Are the compounds the ones you think they are or not? Rules are very different for different classes of compounds and different sources of samples, as Stanley implied.

For those of us interested in prebiotic chemistry some groups of molecules are more interesting than others. The amino acids will certainly come up. We might as well deal with them early. We also need to discuss nucleosides, nucleotides and phosphates. And perhaps sugars along with them. And then there must be other molecules (I can't think of any) that perhaps we might go on to after that.

SIEVER: Porphyrins?

ORGEL: Porphyrins. May I dictate please? We will start by talking about amino acids — how you tell whether they came from where you think they came from, and how you identify them. That may be the wrong order. Maybe we should deal with identification first.

Amino Acids

PONNAMPERUMA: Let's discuss the amino acid questions that have been raised.

MILLER: I'd like to try a gas as a specific example; namely, the cyanoacetylene which you got from the electric discharge.

ORGEL: Well, that will come in *et cetera.*

MILLER: Oh, that's an *et cetera?*

ORGEL: Amino acids, nucleosides, and *et ceteras* — because we must have some order in this, or we will go around indefinitely. I don't mind changing to any other order anyone wants, but I think we have to have an order.

SAGAN: That seems excellent. Especially because I do not have a strong background in chemistry, I would like us to rank the validity of the methods in some quantitative way.

ORGEL: This will be part of the Belmont Manifesto.

SAGAN: Might we find a number which applies to each analytic procedure which describes the probability of making an erroneous identification? For example, mass spectra, because it has many lines and many resolution elements, and a cracking pattern, is obviously much more reliable than something like an amino acid analyzer, or relative retention time data from columns. Can we standardize this? Can we order these methods?

ORGEL: Perhaps Cyril, who has much experience on this, might discuss the problem — which is really one of information content. For example, results from mass spectroscopy yield much more information than a single position on a chromatogram. I will ask Cyril to talk to this (in an informal way, please) and we'll interrupt you as you go along. Cyril, would you please direct your remarks to amino acids?

PONNAMPERUMA: O.K. Firstly, I agree and disagree with John Oró on the two principles he has laid down. Although the system examining can contaminate the system examined in certain cases, if molecules are identified which can never come from the system that is examining there can be no problem. We must look for intrinsic probes. For example, we can have nonprotein amino acids found in electric discharge experiments and in meteorite analysis even when detected in very small quantities which do not come from contamination.

I disagree with Juan's contention that the classical chemical method of identification by three different techniques always works. I'll give an example. Some time ago when we were looking for amino acids in electric discharge experiments I sent out a sample to my colleague who is a very competent analyst. Using gas chromatography and ion exchange methods, he reported the presence of sulfur amino acids.

But, there was no sulfur in the system. He used relative retention times in both cases, and gas chromatography using two different sets of columns.

ORÓ: I'm sorry. You misinterpreted. I think I stated that relative methods per se do not provide any information whatsoever.

ORGEL: John, you can't really mean that. You may not believe the evidence they provide is very powerful, but you couldn't possibly mean they supply no evidence at all. Otherwise we wouldn't use them.

ORÓ: I use them to begin with, but they do not provide any unequivocal information.

SAGAN: But nothing provides unequivocal information. This is why I want a numerical probability attached to each method. Even mass spectroscopy is not unequivocal. You may be bollixed by other compounds which gives a spurious cracking pattern exactly like your mass spectrum. Although unlikely, this has a nonzero probability.

ORGEL: You are asking to quantitate things before —

SAGAN: Certainly they can be quantitated! Why not?

SIEVER: It can't be done. All the possibilities cannot be enumerated.

ORGEL: Carl, you have a very reasonable idea —

SAGAN: So let's spend ten minutes on each method and just estimate numbers for their reliability.

MILLER: It is trickier than that.

ORGEL: It is a tricky business. There are correlations which are difficult to recognize. As far as possible John's three methods should be uncorrelated. The real difficulty is that many problems turn on distinguishing molecules which would be correlated with respect to almost any observation you chose to make. For example, to take an extreme, only optical activity measurements would serve to differentiate the D- and L-forms of amino acids, although they are different compounds, because they are so strongly correlated. I just don't think at the moment we are in a position to place these numerical values on different analytical techniques.

The best we may hope to do (and even this is difficult) is to place a numerical value on a technique with respect to differentiation within a defined class of compounds. We might say: this is a good method for amino acids, but it is a bad method for sugars.

SAGAN: I understand and agree but it seems something much simpler would be more useful than nothing.

ORGEL: But isn't normal intuition (e.g., that the mass spectrometer really does extremely well for this sort of compound, but badly for another) just as good?

SAGAN: No, because of what we have heard here about "unequivocal" and "three different methods." Why three? Why not six or two?

ORGEL: Or one?

SAGAN: Yes. We will never agree unless we put down numbers for each technique.

ORGEL: But since we cannot determine the numbers, there is no point in devising an abstract theory in which these numbers will fit as parameters.

SAGAN: But I was going to say that we can determine some numbers. They are oversimplified, but still many orders of magnitude better than nothing. We can assume there are two chemical compounds, completely decoupled in their properties; that is, they are statistically independent. What is the probability that one compound will fall in the same retention time (or whatever the detection parameter is) as the compound you think you have?

ORGEL: Carl, you must first specify the company from which you bought the fractionator, because that will determine, for example, the width of the peak.

SAGAN: But they do not reflect different orders of magnitude.

ORGEL: Yes, indeed they do! Different breadths of the peak; your parameter thus depends on the manufacturer from whom you bought the analyzer!

SAGAN: How much variance is there?

ORGEL: Enormous!

MILLER: Very high. A factor of ten.

SIEVER: This is a philosophical discussion based on Carl's feeling that lousy statistics are better than no statistics at all. This is not necessarily true, according to some people: the information from "good," "better," "best" is about as good as saying a probability of "0.2," "0.6," "0.8."

SAGAN: Yes, but what if we are concerned with probabilities of 10^{-10} or 10^{-5}?

SIEVER: We say: that is a terrible method!

SAGAN: It's a much finer mesh if we can assign numbers instead of saying good, better, best.

MILLER: Before Cyril got interrupted, he was discussing a case which I thought was of interest. Did the person who returned your "sulfur" compound name the compounds? Did he claim it contained cysteine?

PONNAMPERUMA: Yes, both cysteine and methionine. But there was no sulfur in the reaction mixture! He was simply going on retention times. Here's a case where one needs to combine gas chromatography with the mass spectrometry.

MILLER: How many columns did he run?

PONNAMPERUMA: Two. Two on the gas chromatography, and then a parallel run on the ion exchange.

USHER: Were they very different colums?

PONNAMPERUMA: I believe so.

ORGEL: Cyril, in your experience, if you use retention times on two or three columns, how often would you be mistaken?   And on the other hand, with the mass spectrometer how often would you be mistaken?

PONNAMPERUMA: I'll let Carl put on the numbers.   There's no problem at all with protein hydrolysate.   There are only a certain number of amino acids in proteins and we know them.   Generations of biochemists have run them through columns.   But in the prebiotic analysis, especially in the meteorite and synthetic work, there are many compounds about which we know nothing.   They may have similar retention times to well known compounds.

In our laboratory we run the material through the gas chromatograph and get a retention time.   Then we go for a mass spectrum of those peaks, to try to identify the compound by the mass spectrum.   Then we try to reisolate the compound, or to synthesize it, and reinject it on the column to see whether we get the same retention time.   We come back to the retention time, because occasionally the mass spectrum may not give you an absolutely firm identification.

ORGEL: Now, we won't use the evidence to incriminate anyone, but how often do people identify the compounds by mass spectrometry, and then find the compounds have been incorrectly identified?   Is that a thing that never happens any more?

PONNAMPERUMA: Mistakes might be made with regard to an isomer, maybe, but, generally, you know, if you identify a particular amino acid, I think, or a heterocycle, probably it is a good bet that you found it there.

ORGEL: A good bet!   Would you offer me ten-to-one, or 100-to-one, or what?

SAGAN: Watch it, you're getting quantitative!

PONNAMPERUMA: There is a class of compounds I would say the bet is 100-1.

FERRIS: I have an example – a very recent sad example, of identifying a compound we isolated.   We had the proper mass spectrum with an authentic sample run on the same machine, and we decided we'd just tidy up by checking on thin layer chromatography.   The compounds were not the same.

LEMMON: Does this happen one time out of a hundred?

FERRIS: This is a rare experience.

MILLER: Could you be specific?

FERRIS: We thought the compound was orotic acid, obtained from hydrogen cyanide.   We derivatized and prepared a butyl ester of orotic acid.   It seemed to have many of the properties of orotic acid, but on more detailed analysis – namely, thin layer chromatography – it wasn't.

USHER: Do you know what it is yet?

FERRIS: No, we don't.

PONNAMPERUMA: Did the mass spectrum show you the presence of a six member heterocyclic ring?

FERRIS: We ran authentic orotic acid alone. The mass spectrum of the compound showed all the peaks were there. The variations in intensities were slight and within our experience. Put side by side with orotic acid you would say this was the same material.

ORÓ: Was this in the probe?

FERRIS: Direct inlet.

ORÓ: You actually made the analysis. Could it have been a product of decomposition of orotic acid? Depending on the volatility of the compound, you may —

FERRIS: Are you suggesting it may be rearranging in the mass spectrometer? Sure. It could, but I'd be very surprised if it is not very closely related to orotic acid.

USHER: What probe temperature did you use?

FERRIS: I'm not sure.

ORÓ: For whatever it may be worth, in our experience mass spec has given us no problems so far.

ORGEL: What is the experience of the group? Does anyone disagree with Juan? By and large, with only rare exceptions is a mass spectrometric identification acceptable?

USHER: Should we distinguish high resolution from low resolution?

ORGEL: We are now talking of low resolution.

PONNAMPERUMA: Certainly high resolution gives a much more refined picture.

USHER: I have done much high resolution work, and can think of one example where microanalysis gave an indication of something which turned out to be what looked like the parent peak on this particular compound at low resolution. On high resolution, however, it was a different compound, with a carboxylate group which was not seen in the parent peak. That had come off, as usual. This would not have been solved without high resolution.

PONNAMPERUMA: In case of doubt we ought to do it by high resolution, which is generally what we have done.

SAGAN: Is the improvement three orders of magnitude, or so?

USHER: The RMH-2 gives you up to 100,000 resolutions but —

SAGAN: Compared to what for the low?

USHER: Unit mass of about 300.

SAGAN: This again confirms my feeling that we should do this quantitatively. Since here you get several orders of magnitude in resolution you statistically decrease the probability of a false positive by those several orders of magnitude.

USHER: I agree with Cyril that it has to be combined with the purification –

PONNAMPERUMA: This is what Juan said. If the separation of your material has been made by gas chromatography, the mass spectrometry becomes far more meaningful. One might even need an earlier step: an ion exchange system, then a gas chromatography, and then a mass spectrum, because gas chromatography alone has given problems. For example, it doesn't show up as peaks; you don't see the compound on your gas chromatograph but it appears in your mass spectrum because the column doesn't separate it, but the carrier gas takes it through.

ORGEL: May we say, then, that a high resolution mass spectrum is the ideal? Most of the time low resolution mass spectrum suffices, but occasionally we have troubles and require high resolution. Is this the consensus? Suppose either the materials or quantities preclude mass spec analysis? Are there other acceptable methods of identification? Are data based on retention times ever acceptable? (Please remember we're still talking about amino acids in prebiotic experiments).

BADA: I encountered a problem on the GC-MS in a couple of cases. We had a particular peak that gave a very funny mass spectra. What we collected that peak off the GC, made some Dansyl derivatives, or some DNP (dinitrophenol) derivatives, and found in fact, three different compounds eluted at that same peak.

If you just start taking mass spectra of these peaks how do you determine that you don't have a mixture of compounds? I have found mass spectra, when there may be a mixture, are extremely difficult to interpret.

ORÓ: A possible solution is coming up now, it is called mass chromatography. It's a little more involved than ordinary GC-MS. Your instrument must be used to scan a single ion peak.

I can't give you details, but Dr. Hammar from Sweden just gave us a seminar and, indeed, we used this method to characterize some of the components in the lunar sample. The instrument is tuned to a certain ion, the chromatogram is scanned for that particular ion, and then the results are plotted. It gives an equivalent gas chromatogram, but rather than being measured by a gas chromatogram, the detector is measured by the distribution of an ion in the GC, giving what we call "mass chromatography."

What appears to be a single peak may actually be three or four peaks.

And then, of course, the problem is that you may want to scan for more than one single ion — for ten ions, let's say — which gives you all the sophisticated detail of the three compounds. The sophistication in gas chromatographic methods will make possible the determination of cases where single gas chromatographic peaks contain a number of compounds.

USHER: This depends on the compounds eluted. If we have two compounds, one at the leading edge, and the other at the tail — if the whole thing had been exactly coincident, and gone to completely symmetrical peaks — which is certainly a fortuitous event — this method would not have helped.

The method that is used already is to scan rapidly over a huge mass range, both at the starting edge and in the middle (in fact, you can take ten cuts throughout a single gas chromatographic peak). But if all three compounds are coming out, starting at the same time, peaking at the same time, and finishing at the same time, this will not help. Is that what happened?

BADA: I think so. We took five or six mass spectra across that peak, and there was a little difference at the very first, but other than that they were all the same, and it was — the cracking pattern couldn't be fit to anything.

But working with the Dansyl and DNP derivatives, which we then put back on the GC analysis, we could interpret it.

USHER: So even high resolution mass spectroscopy is not completely foolproof.

ORGEL: Let us now discuss what must be the major method in use; namely, of retention times. Are there any circumstances under which an identification can be made without a mass spectrometer, on the basis of retention times?

SAGAN: Let's ask: How many different, non-mass spectrometric methods, simultaneously applied, give a reliable identification?

ORÓ: May I discuss this now, and provide some practical guidelines? First, I would use a relative method that applies whether it is the ion exchange, GC, thin layer chromatography, or any other. Then I use three other separate methods (four is equally good, ten is better).

After the relative method, I need either UV, infrared, or some other common method that provides additional information. If there are ambiguities, I go to something with the high degree of probability (in Carl's terms) that the compound is the compound. I'm biased for selecting the combination of gas chromatography and mass spectroscopy because it has the synergism of two methods that adds additional certainty.

And then, a third method, either a derivation, a melting point, or other more specific methods. With the combination of these four methods, I think the data should be accepted.

ORGEL: This is getting to be a motherhood discussion. Everyone agrees by now what the maximal requirements are: all available methods! However we need a completely different type of conclusion: what are the minimal methods of identification? We can agree on a really outstandingly good piece of work, but how do we decide whether or not to publish our results. If the "aspartic acid" moves to roughly the right place, how do our experts decide whether it really has been identified as aspartic acid? What's the shoddiest job that still identifies an amino acid?

People seem resistant about expressing views on this.

MILLER: Let's take a specific example. Even though melting point determinations are old-fashioned, I think they are very effective, yet melting point alone does not identify a compound. But a melting point and a mixed melting point should be extremely reliable.

FERRIS: But you usually can't get enough material for a melting point.

ORGEL: I feel an extraordinary resistance to discuss this classical method which all of us have used at one time or another; namely, identifications which are based mainly on retention times in either paper chromatography or on columns. Will someone say specifically that they are (or are not) reliable?

MILLER: I don't think ten paper chromatographic solvents are necessarily any better than two or three.

ORÓ: It is! I have a case of a compound that behaved identically with four solvents and was distinguished with the fifth.

USHER: I have such an example with thirteen solvents!

ORGEL: We also had one with thirteen or more solvents.

MILLER: Then you must proceed to a different basis of separation, for example the GC. But several solvents and the GC still leaves me uneasy.

USHER: Electrophoresis is another, of course.

MILLER: Right. That is a separation based on a different property, but I would still feel uneasy about that.

USHER: You can do electrophoresis at different pH's, for instance.

MILLER: But you will get wiped out by alpha cytidine!

ORGEL: Yes. That's an awful problem, but let's stay with the amino acids for the moment.

FERRIS: May I suggest what I think is a minimum method, and let people jump on me?

We're discussing amino acids. The usual procedure is the identification on an amino acid analyzer. The technique from there is that of a retention time on an amino acid analyzer: the conventional injecting the same material in, to see if your particular thing still gives you a symmetrical peak.

The next step is an elution of the material: collect it from the amino acid analyzer, and do chromatography in two or three different solvent systems. If it has the same R_f values on two or three other systems, you would have done an ion exchange chromatography and a paper chromatography. Although this is not the most elaborate analysis, it is reasonable for an identification.

BADA: How about a synthesizing derivative? We collect the peaks that are eluting off the amino acid analyzer, make a derivative, and then do paper chromatography. This fits your minimal criteria even better, because the derivatization step is a very easy thing if you are making dansyl derivatives. Just add dansyl chloride and then if you do the paper or thin layer chromatography on these, it might be a minimal criterion for identification.

MILLER: Would you be content with taking the peak off the amino acid analyzer, chromatographing this peak on paper with several solvents and getting values from paper for an identification?

FERRIS: Yes, this is what I'm saying.

MILLER: Although that is pretty good, it is not publishable by my standards.

ORÓ: That would not be enough for me either.

NAGYVARY: Chemists might not like it, but what about the identification of amino acids by biochemical techniques? With protein synthesis, which has worked out beautifully, you can get purified ribosomes — use a system with that particular amino acid missing. The amino acid activating enzyme technique allows the synthesis of a specific viral protein, for example, which can be measured. Then your amino acid belongs in the natural series.—

ORGEL: That's interesting, but there are dangers in these techniques. First, biological systems behave in a defined way with the familiar amino acids, but in prebiotic experiments there is always the possibility that you have something else. Since the biological systems never have seen them, we really don't know what to anticipate.

NAGYVARY: This is not an exclusively good method, but it is worthwhile to try with small quantities.

ORGEL: Since we were getting close to the nub of the matter on the method of analysis, it would be a pity to let it escape!

WOLMAN: I feel uneasy about using the amino acid analyzer, isolating the peak, and then running up another chromatogram or another electrophoresis. First I worry about the ratios at 440 to 570 mμ on the amino acid analyzer. For example we had a compound which we were quite sure was not glutamic acid. Although it comes where glutamic acid does, the ratio of the peaks is completely different. Where usually the ratio is about one-to-three with glutamic acid, we got about one-to-seven.

After taking down the peak and eluting out the substance from the amino acid analyzer, I would derivatize the compound, run it on the GC-MS, and compare up the results with the known amino acid and then I would feel more easy.

ORGEL: You would insist on the mass spec?

WOLMAN: Yes, I would be compeled to insist on MS. And furthermore I want to work with the right derivatives. I must show that I have a D-L compound and not an L compound. Contamination, presumably, would be mainly of L-amino acid.

FERRIS: This seems to exclude anyone who does not have a GC-MS.

ORGEL: Yes. Does this mean no work on amino acid prebiotic chemistry can be carried out by those who lack a GC-MS?

USHER: No reliable work.

ORGEL: No publishable work?

ORÓ: The samples should be sent to someone else who has it.

ORGEL: This is not a joke. Are we really concluding that no work should be published on identification of amino acids unless the man has, or has a friend with, a GC-MS?

USHER: Since you can't claim that you need too much material for GC-MS, I think we must agree if we are going to do good work.

MILLER: I think it is a bad idea to state categorically that no work is publishable. If the person does the GC and the wet columns and the paper chromatography and the paper electrophoresis, he has already done more than was done in 90 per cent of the papers in the prebiotic synthesis literature.

ORGEL: This is why I raised the question and pressed it.

MILLER: I don't know whether our standard should be raised that high so quickly!

SIEVER: Perhaps our manifesto could simply state that this kind of data must be verified by those with a GC-MS to be published.

USHER: Stanley, we may raise the standard now because the equipment is available.

MILLER: Before we try to boycott or suppress papers, (as was implied, I think, by Leslie's statement) I think we must be a little careful. It's fair enough to say that using only paper chromatography or only the amino acid analyzer is inadequate and good work ought to have the GC-MS.

MILLER: The problems are with the in-between cases. Suppose somebody is doing some experiments in Italy, where he simply can not get to the GC-MS.

USHER: No. He still should be able to, because NIH runs a service. Anyone can send samples in to Fred McLaughlin and he'll run them.

SAGAN: We have the convenience of actually being at Cornell, and still our samples are not very rapidly run. This discussion is puzzling. Let us compare the very admirable standards of reliability talked about here with standards deemed reliable in other areas of science. Let's take an aspect of my business for example: spectroscopic identification of compounds in planetary atmospheres. There is a large literature on this subject. A fair number of compounds that have been identified have turned out to have been spuriously identified. I guess about five or ten per cent of all compounds published as having been found in a planetary atmosphere we now think are spurious. That does not mean we ignore the 90 or 95 per cent we think are good, because there is various supporting evidence for them.

If I had a technique to guarantee a 99 per cent probability of successful identification of compounds in planetary atmospheres I would be delighted. You ought to have seen the tiny little peaks on a few spectrograms for the first identification of water in the martian atmosphere.

By our standards I believe you fellows are rich.

ORGEL: Everyone would be delighted with a reliability of 99 per cent, but I doubt whether the identifications in the literature in the prebiotic chemistry will approach 99.

SAGAN: Okay. From my point of view it still would be very interesting to know the number that characterizes the reliability in paper chromatography, in gas chromatography, and all the column techniques. Are these independent methods? If so, when you apply two or three methods, you multiply probabilities. If each technique has a probability of being wrong of 10^{-1}, and three independent techniques are used, the probability of being wrong on all three is 10^{-3}, which is very respectable.

PONNAMPERUMA: Carl's point of view suggests that many more prebiotic chemistry papers will be accepted for publication.

ORGEL: I'm only pushing this as a sort of joke, still it seems necessary to discuss our standards.

PONNAMPERUMA: I believe a lot depends on the compound identified. If somebody has identified 10 microgms/gm of glycine by ion exchange chromatography, separated it out and has got one spot, we would be inclined to accept that identification. This may not be the same for a compound that doesn't normally come into laboratory practice. Much depends on both the compound and the quantity.

MARGULIS: Representing the poor relatives that lack GC-MS, how fast can we get an agreement concerning the status of the sulfur containing amino acids? Is there a consensus in this group?

There are reports in *Biochemical Predestination* (Steinman and Kenyon 1970).

SAGAN: And methionine.

MARGULIS: Yes. Does this group agree that you can get cysteine and methionine under prebiotic conditions?

PONNAMPERUMA: I'll turn this over to Dick Lemmon.

LEMMON: I think there's no question about this, because it involves another extremely reliable method. I would assert that it perhaps approaches and may even exceed mass spectrometry as an identification method. In this case it involved carbon[14] labeled methane. The sulfur amino acids were then identified, they had carbon[14] in them, and formed on the chromatogram. An x-ray film is used in the dark room, and you get a radioactive spot. Then to that radioactive material you add known carrier methionine. Then you rechromatograph on the paper, and now you can spray with ninhydrin, getting an ninhydrin colored spot, and here you have the same R_f value for the radioactivity and for the color, but in addition, the very outline of the color and the radioactivity (all the bays and peninsulas, if I can put it that way) are identical. This is vastly different than having two comparable R_f values –

ORGEL: I ask innocently, but is this really true? Aren't the bays and peninsulas due to the way the solvent is flowing, and –

LEMMON: But why do the radioactivity and the color have exactly the same outline?

ORGEL: Might you have only one parameter: namely, the center of the thing – because the shape doesn't matter, because that's the way the solvent is flowing through? Isn't that possible?

PONNAMPERUMA: Let me answer. Having used that technique quite extensively, I would say that in 99.9 cases, it works out.

LEMMON: I don't believe you know of one in a 1000 exception, in which the entire outline was identical.

PONNAMPERUMA: This is what happened.

MILLER: What is the example?

ORGEL: Nucleoside.

PONNAMPERUMA: Yes. That's where the technique failed.

LEMMON: I don't believe the patterns were identical in all respects.

ORGEL: We had an awful job of separating them by other methods.

LEMMON: Anyway, we're down to one chance in a 1000 here. I'll accept that.

PONNAMPERUMA: Let's come back to the methionine. I don't know. There was some reference to amino acids synthesized from ammonium thiocyanate. You are referring to the one in which you introduced C^{14} labeled methane and used the electron beam as a source of energy.

LEMMON: Yes.

PONNAMPERUMA: You see, there are two identifications of sulfur-containing amino acids, the one that you [Lemmon] published earlier (Choughuley and Lemmon, 1966), and the other one which was published by Steinman later (Steinman et al., 1968). These are the two cases.

SAGAN: There's a third. We have a paper in press now on the sulfur containing amino acids (Sagan and Khare, 1971; Khare and Sagan, 1971).

PONNAMPERUMA: Yes. I am aware of that.

LEMMON: The first two reports were both based on the same fundamental procedure, the carbon[14] labeled methane.

PONNAMPERUMA: Yours was. I'm not sure about the other one.

MARGULIS: For us out of the field, that's very important.

ORGEL: I'd like to finish up this discussion, just seeing how much time we have already spent only on amino acids.

Carl, have you anything to say just briefly about your sulfur work?

SAGAN: Yes. We have three different techniques which give cystine and/or cysteine, but not methionine. We use methane, ammonia, water, and H_2S as the photon acceptor. We photodissociate the H_2S with a long wavelength UV source, hot hydrogens come off, break up other things, and in recombination make amino acids.

There probably won't be any question about alanine and glycine, and so on, but we also claim to have made cystine and cysteine. The identification is based upon labeled carbon (the methane is $C^{14}H_4$) and we have autoradiography in two different solvent systems.

We also have it on the amino acid analyzer and now in gas chromatography. We obviously will try the GC-MS as well. We are trying to respond to the small probability of being in error, but if I can assume these techniques are even a little bit independent, and look at the probability of being fooled on each one of them, and multiply the probabilities together, I conclude that the likelihood of being wrong is smaller than the likelihood in most planetary spectroscopy experiments.

This makes me quite relaxed.

MILLER: But the distinction between the planetary spectroscopy and prebiotic chemistry is that astronomers do not have the techniques really to do very much better, whereas prebiotic chemistry identifications can be done very reliably if you are willing to put in the work. The planetary spectroscopists can be excused.

SAGAN: No, that's not really right. A Fourier transformer spectrometer, although exceedingly difficult to use, can be used to get much higher resolution. This can be done in planetary spectroscopy.

USHER: How much material do you have?

SAGAN: Actually not too little, about 0.04 micromole per cc in some 10 cc.

USHER: Because, Leslie, one technique which will be used for this, if there is enough material is C^{13} NMR. I would accept this as readily as high resolution mass spec.

We come back again to how much material can be made.

ORGEL: Obviously, we can't go on in this detail, but I would like to discuss contamination in amino acid analysis. Let me tell you the things I always worry about, and see whether there is reason to worry.

First when dealing with small quantities there is the obvious contamination: fingers, dust, and so on. Secondly, there is the quality of reagents: how much amino acid is hidden away in a bottle of HCl? Clearly, these things are only important when you are concerned with parts per million. Let's discuss now contaminants from the investigator and equipment, and then contaminants from the reagents. Stanley?

MILLER: I think Dr. Wolman can tell us about this.

WOLMAN: We were interested in checking up the contamination of hydrochloric acid. We found that newly opened bottles of concentrated hydrochloric acid contained between 5 to 8 x 10^{-6} moles amino acid per liter.

LEMMON: Would you give me those numbers again?

SAGAN: Right! Say it two or three times.

WOLMAN: 5 to 8 x 10^{-6} moles amino acids per liter.

SAGAN: Would you specify the brands?

WOLMAN: We checked two bottles: a Baker analyzed, and an Allied Chemicals.

SAGAN: And what was the normality of the HCl?

WOLMAN: It was twelfth normal, concentrated.

MILLER: 37 per cent.

BADA: Which amino acid was most abundant? Might you give us an idea of the relative proportions?

WOLMAN: There was plenty of glycine and alanine. There were some amino acids which came off near aspartic acid but the funny thing was we had high 440 and the low 570, which indicated some kind of an amino acid.

PONNAMPERUMA: Amino diacetic.

WOLMAN: Maybe. There were few spots in the area of valine, isoleucine and leucine, and some traces of basic amino acids.

BADA: How about serine?

WOLMAN: There was no serine, but this does not mean much, because serine is partially destroyed under acidic conditions. Maybe serine was there which was destroyed before. There were some traces of threonine.

At that point we decided to check our constant boiling hydrochloric acid, which we made by mixing the concentrated hydrochloric acid with glass distilled water one-to-one, and distilled it. We found that the constant boiling acid contained somewhere between 2 to 8 x 10^{-8} moles of amino acids per liter. This amount might not cause too much trouble in organic synthesis work. But in looking for amino acids in sediments, meteorites, lunar samples, or any other samples containing very small amounts of amino acids this might make a big difference. We thought about the possibility that some of the amino acids just volatilized and distilled during the distillation of the constant boiling HCl. We took a distilled constant boiling hydrochloric acid – and added to it one micromole per millimeter of glycine, alanine, valine, lycine, and glutamic acid. We distilled the constant boiling hydrochloric acid, and got some very interesting results.

MARGULIS: Were these racemic mixtures?

WOLMAN: These were L-amino acids which were taken from the bottle.

[Slide] Here we have our distillation plot. We were using two Kjeldahl bulbs at the top of the distillation flask, in order to avoid any spray. Here we have the ratio of the amino acid concentrations in the distillate to the amino acids in the pot in various times.

We really get a big increase of amino acids once we are down to about 75 per cent distilled. The most volatile is valine, then alanine and glycine; glutamic acid and lysine are much less volatile.

Even in the region of 20-60% distillation we still have a small amount distilled, but relatively much less amino acid than at the end.

It is interesting that we are looking at the standard procedure for making constant boiling hydrochloric acid! This procedure calls for discharging the first three-quarters and collecting the part between 75 and 95 per cent, which is the part containing all the amino acids!

Of course, using this, we contaminate ourselves. We should better use the area between 20 and 60. We were able to get the equivalent of constant boiling hydrochloric acid by using Matheson electronic grade hydrochloric acid, collecting it in liquid nitrogen, weighing out, and transferring the known amount into the required glass distilled water. This contains less than 0.2 nanomoles of amino acid per liter — or at least 0.2 nm is our limit of detection. We may be getting even less.

The amino acids we think are in those 0.2 nanomoles are some traces of aspartic acid and glycine. I think from now on when searching for amino acids in earth sediments. lunar samples, meteorites, or in any source where we suspect only small quantities, we have to make our own constant boiling hydrochloric acid.

ORGEL: May we have comments on this? Please address yourself to the problem of contaminants in reagents.

PONNAMPERUMA: Six years ago we had exactly this experience. Bill Schopf was looking for amino acids in the Fig Tree chert in our lab. We discovered analytical reagent HCl and ammonium hydroxide both had amino acids.

ORGEL: Please tell us about the ammonia.

PONNAMPERUMA: Amino acids must come from little bugs that at some stage in the course of the preparation get into the system. The common amino acids found were glycine, alanine, aspartic and glutamic acid, serine, and threonine. This contamination can be overcome only by using triple glass distilled water, combined with ammonia gas, to prepare ammonium hydroxide and HCl gas combined without tripled distilled water. We have gotten less than 5 nanograms, about a 20th of a nanomole. This is the level at which we are now working at the lunar laboratory.

Another comment on contamination, lets switch from reagents to fingerprints (I know Juan Oró and Bob Anderson have published their fingerprints!) I have a gas chromatographic fingerprint of Charles Gehrke.

LEMMON: Does the FBI have it?

PONNAMPERUMA: Fig. 5 shows the N-TFA and butyl ester standards below; and Gehrke's fingerprint. There are a number of peaks: alanine, valine, glycine, leucine, isoleucine, proline, aspartic acid, phenylalanine, an unidentified amino acid, presumably glutamic acid, and something else.

Clearly one has something to be extremely careful about in handling material.

SAGAN: Perhaps you can explain something. There are several reports of the production of compounds under presumed prebiological conditions, where after acid hydrolysis the amount of amino acid goes up by an order of magnitude or more. Cyril, (or Dr. Wolman) do you think that the increase is not because bonds in prebiotic polymers are broken but because people are introducing free amino acids?

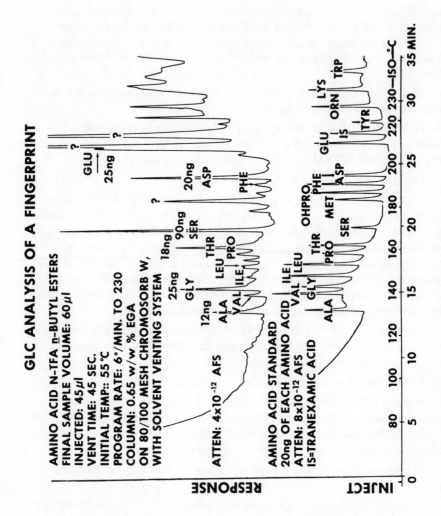

Figure 5. Gas-liquid chromatographic analysis of organic material from a single individual's fingerprint.

PONNAMPERUMA: It depends on quantities. In the work you refer to there is a dramatic increase in the micrograms per gram quantities. This might involve a small contribution from the acid but I think most of it is from either hydrolysis of nitriles to amino acids or the breakdown of some kind of polymer. However, in the nanogram range, I would be very suspicious.

ORGEL: We must be peculiarly suspicious of compounds which may be difficult to make on a prebiological mechanism.

SAGAN: You say there is a qualitative test, and Cyril says there's a quantitative test.

LEMMON: I'm confusing the prebiotic synthesis discussion with looking for compounds in meteorites. Again I just want to emphasize the use of carbon[14] in prebiotic synthesis experiments of course as a substrate. This avoids a lot of trouble. Although this doesn't apply at all to meteorite work et cetera, those who don't want to buy a mass spectrometer might consider using carbon[14] as a substrate.

SAGAN: But it can't be used in the mass spectrometer.

LEMMON: No. You don't need a mass spectrometer. It becomes less necessary to have one.

BADA: From this data we can more or less set a limit on the amount of detectable amino acids using certain reagents. If somebody is using concentrated HCl to detect quantities of amino acids on the order of 100 nanomoles or so, it is obvious they are studying contamination only. If just one distillation is used to improve your concentration, I think using 100 ml for hydrolysis, that on the order of, maybe, 20 to 50 nanomoles is the limit. With really low concentrations – tenths of nanomoles – you must use Cyril's method of making your HCl right from gaseous HCl.

PONNAMPERUMA: And triple distilled water.

BADA: Is triple distilled necessary? I have found that double distillation in a filtration is usually good enough.

MILLER: Does the third distillation really make a difference?

PONNAMPERUMA: Yes, this is our experience. Maybe your second distillation is a much more efficient distillation than our third one.

MILLER: Do you use alkaline permanganate when you distill it?

PONNAMPERUMA: No, we don't use any permanganate.

MILLER: The old-fashioned way, which I think is very effective, is to use alkaline permanganate in the water being distilled. It is said to oxidize all the organics.

ORO: There is the added problem that the acid hydrolysis applied before the analysis may bring artifacts, that is to say, amino acids that are neither present as such in the geological samples analyzed, nor have they been synthesized in experiments under prebiotic conditions. It is well known that sugars and other compounds in the presence of ammonium salts or ammonium derivatives rearrange and produce amino acids. I think Dr. Miller has worked on this. In our first paper on the synthesis of amino acids (Oró *et al.*, 1959) we studied, among other things, the action of HCl on a mixture of formaldehyde and hydroxylamine. About an order of magnitude more amino acids were formed under acidic conditions than under basic conditions.

ORGEL: Let us break for ten or fifteen minutes for coffee.

Nucleic Acid Derivatives

ORGEL: Next I'd like us to do the same sort of job on nucleosides and nucleotides. The techniques here are really rather different. Mass spectrometry to date has not been used extensively, although some may suggest it should be. Chromatography, ultraviolet spectroscopy, and enzymatic identifications have been, perhaps, the main methods for differentiating isomers amongst the nucleosides and nucleotides.

The problem of contaminants is perhaps more serious than in the amino acids because most commercial biochemicals are obtained from biological sources, and contain miscellaneous biological junk, including often the compounds you are trying to make.

Now, where shall we start? Bob, would you like to say something about identification to begin our discussion?

SANCHEZ: Let me begin with the specific example of beta-cytidine. We found a synthesis that seemed to produce yields of approximately a tenth of a per cent. We found perfect coincidence of R_f's in about fifteen chromatographic systems — descending systems with all types of solvents and electrophoretic systems. The coincidence between our compound and β-cytidine was very good. At first glance, the ultraviolet spectra of our unknown and of authentic beta-cytidine were identical. Toward the shorter wave length region there were differences in the spectra, which we attributed to some minor contamination, and ignored.

In all attempts to further purify this material, that minor difference in the UV spectra persisted. We eventually tried a chromatographic system described by Charles Dekker (1965) which separates ribonucleosides on a strongly basic anion exchange resin. There the separation of the two compounds was like night and day. It became very clear that we had alpha-cytidine, not beta-cytidine, which is very different.

ORGEL: I'd like another cautionary text. Cyril, would you please talk about the corresponding difficulties in the deoxy- work?

PONNAMPERUMA: This concerns the synthesis of nucleotides, and is

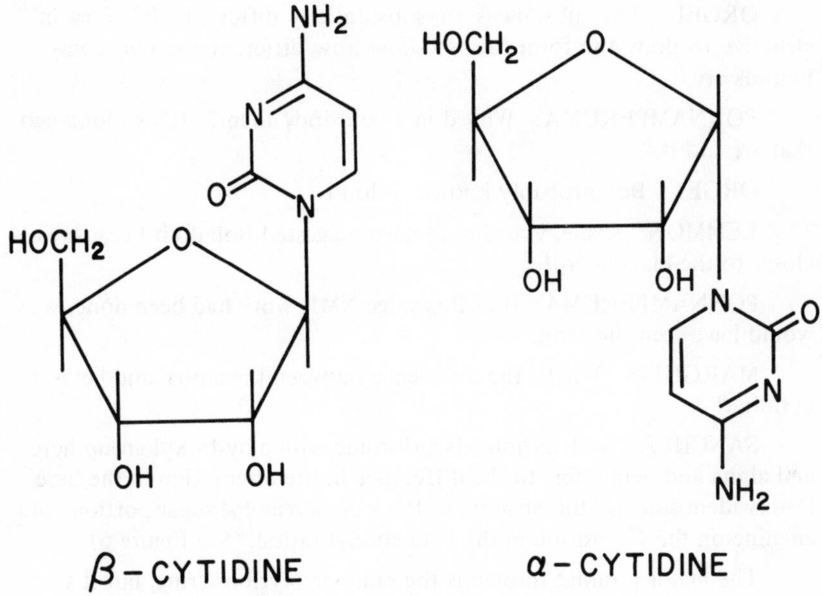

Figure 6. Structures of alpha and beta cytidine.

apropos of the remark that Dick Lemmon made. I worked in the same laboratory at one time in Berkeley, and one technique we swore by was the superimposition of the autoradiograph over the "shadowgram." This is produced by taking a UV photograph of the chromatogram itself. This gives you a white spot on the shadowgram, and a dark spot on the autoradiogram. When superimposed the outline of the shapes match beautifully and precisely in every detail.

We had done the adenine synthesis earlier with methane, ammonia, and water and had the evidence in this manner. In the case of the nucleosides we initially obtained the result with three different chromatographic systems, separated on methane and water. We extracted the spot out and ran it on a two-dimensional system. I don't recall the solvents but precise coincidence was observed.

We had yields of about 7 per cent. We felt so confident that we published a note in *Nature* (Ponnamperuma and Kirk, 1964). Later when we examined this, however, with Professor Reid we found that this yield didn't hold. The conclusion was that there was probably less than 0.3 per cent of the right isomer there.

ORGEL: The substances are substantially different. It's very instructive to draw the formulae, to show how different the two compounds are.

PONNAMPERUMA: Why don't you draw them? It's so long ago that we did it.

ORGEL: Bob probably knows. I don't.

LEMMON: Leslie, you should have suggested Bob do it because he's closer to the blackboard!

PONNAMPERUMA: If at this stage NMR work had been done, we would have seen the error.

MARGULIS: What's the difference between the alpha- and beta-cytidine?

SANCHEZ: Well, cytidine is a riboside with a hydroxyl group here, and alpha and beta refers to the difference in the orientation in the base. Deoxyadenosine has this structure. It's a deoxyribosyl sugar portion, and adenine on the 1' position in the beta configuration. (See Figure 6)

The major product formed is the same deoxyribose ring, but it's in the pyranose configuration, which is a 6-membered ring. The adenine is linked not to the 1' but to the 3' position. It's not a nonreducing sugar, as deoxyadenosine, but it's still a reducing sugar with the 1' position free.

However, chromatographically —

PONNAMPERUMA: Identical. One point that enabled us (that is, Leslie and his colleagues) to distinguish was the acid hydrolysis. One compound is stable to acid, the other is not. We missed that point in our controls.

Incidentally, the mass spectrum would not have told us much on this.

FERRIS: I'm not sure. I had the recollection also that you felt you had some phosphate derivatives as well, Wasn't it in this system that the fire-fly test gave you —

PONNAMPERUMA: No. No. The phosphate experiments were done separately, where a nucleoside was phosphorylated. There we had substantial evidence; we were starting with a nucleoside, and we were able to distinguish the isomers, the 2', the 3', the 5'. We had not only the coincidence, but also the ratios of carbon[14] and P[32]. So there was no difficulty.

NAGYVARY: As a prebiotic chemist, I feel there is no excuse whatsoever in 1971 for not identifying these compounds by NMR. In the case of alpha and beta nucleosides, with NMR the beta nucleosides give a 3+, for example, for the 1' component, and the ortho-nucleosides for the water.

We had some thio compounds, and without NMR it would have been impossible to tell the exact structures. In some cases the NMR is also not enough.

ORGEL: That was actually my next point. I'd like some examples where, even with enough material and the machine to do the measurements, you may still go wrong on the basis of NMR tests.

NAGYVARY: We have new compounds which are very hard to interpret on NMR.

ORGEL: I would like to discuss a different question — not the determination of structure but the identification of compounds by the NMR. Is it possible to believe a compound is beta-adenosine from its NMR, but, in fact, it is something else? Do you have examples of that?

NAGYVARY: I think NMR is an enormously powerful tool in identifying all the natural nucleosides. I think NMR gives 99 per cent accuracy. If its chromatogram and UV spectrum look like adenosine, and if the NMR is identical, you can account for every proton in a very characteristic position.

ORÓ: What quantities do you need?

NAGYVARY: It depends on your instrument. If you have a cath computer, with 5 mg of adenosine you can run 50 runs —

PONNAMPERUMA: Can't you use much less, say half a milligram?

NAGYVARY: If you are a synthetic chemist, you must be able to produce 5 mg! We are not analyzing moon samples, but doing an adenosine synthesis!

USHER: You can use as little as a hundredth of a milligram using preocanthin.

ORGEL: What are your opinions concerning the ORD technique? Is it useful?

NAGYVARY: For the identification of natural nucleosides, it is fairly good, but if you have some sort of deviation, you can not be sure. In making a differentiation between alpha and beta and between new derivatives, I don't trust the ORD very much.

ORGEL: Another major topic is the use of enzymatic methods. This may be a most powerful identification method. I'm now referring to material available only at optical density levels. Adenosine can be deaminated to inosine, and the change very readily shown chromatographically.

Do people have views on (a) its discrimination — reliability, in that other compounds do not interfere — and then (b) its sensitivity?

LOHRMANN: We use enzymatic methods extensively, in combination with chromatographic and electrophoretic methods, to characterize the phosphorylated nucleosides.

For example, there is no way, as far as I know, to distinguish chromatographically or electrophoretically between a 5' and a 3' phosphate of thymidine. The only method is by using an enzyme. We used this crude renin, which dephosphorylates the 5' but not the 3' and just by combination of that we get the ratio of 5' to 3'.

In the ribo- case we distinguish them mainly by borate complexing. The cis hydroxyls in the presence of borate run chromatographically different from the 3' and 2' phosphates.

USHER: Would P^{31} NMR cause any difference in the byproducts?

LOHRMANN: We have not used NMR for this, because I don't need it.

MILLER: Is there any possibility errors are made?

LOHRMANN: I think there is definitely no error possible in distinguishing the 5' and 3' or 2' phosphates when the borate system is used, because of the big difference.

PONNAMPERUMA: The 2', 3' and 5' can also be separated fairly easily on ion exchange systems.

LOHRMANN: Yes.

PONNAMPERUMA: Leslie, if you allow me, I'd like to say something about the enzymatic analysis which should have been said in the amino acid discussion, about the use of LSP (the leusoamino pepsidase) an enzyme to break down peptides. This is a classical test used by biochemists. We have tried this to see whether peptides are present in some prebiotic mixtures. There are horrendous problems, because LSP always leaks a few amino acids, and we don't know where they come from.

ORGEL: Perhaps we should talk about the different modes of linkages; but maybe that's too specialized. Do people want to talk about distinguishing the 2'-3', 3'-5'.

SAGAN: May I pose an alternative? Bob Sanchez has attempted to put some quantifications on the reliability of different techniques. May we hear that discussion?

PONNAMPERUMA: I think one can make a general statement: for nucleosides and nucleotides paper chromatography alone, or thin layer chromatography alone, is very poor. One needs NMR studies.

ORGEL: Another general question is: to what extent was what you make in these nucleotide syntheses present in the mixtures from which you start?

We have somewhat horrifying experiences along these lines. All ribose contains a per cent or so of adenosine, is that correct?

SANCHEZ: It varies. A bad example is ribose-5-phosphate, which is contaminated with as much as a few tenths of a per cent of adenylic acid.

USHER: Are you speaking of commercially available material?

SANCHEZ: Yes.

USHER: If you make it yourself —

SANCHEZ: Yes. If you use commercial sources in a prebiological experiment, you get very nice yields of nucleotides.

USHER: We are gradually discovering that we need to set up our own chemical company to do these experiments.

SANCHEZ: Or take the precaution of analyzing everything used which can be very troublesome.

USHER: And each time it is used, because people change their methods.

SANCHEZ: Yes.

ORGEL: It's quite interesting to recount how we discovered this. After a very long period of being unable to make adenosine, an adenosine synthesis was discovered that worked very well. Since subsequently all experiments of every type worked very well, we concluded that no matter what we did, 0.7 per cent of adenosine would be produced. This was a great compliment to the analytical techniques of my colleagues, because they always found this 0.7 per cent. Then, we went back and discovered, indeed, the starting ribose phosphate contained just this amount of adenosine!

PONNAMPERUMA: Every chemical bought commercially must be examined. This is a rule of thumb. Gas chromatography can often help here, for example with derivatives of sugars, sugar phosphates, to detect the presence of small impurities which otherwise will not be seen.

USHER: If you are interested in absolute purity, there's no guarantee that the impurity will be seen in the method you use. If it's involatile, you know you have got mostly what you want.

PONNAMPERUMA: Gas chromatography is very useful technique for knowing what you are dealing with.

USHER: You can tell you have got what you want, but you can't tell if there is something that isn't appearing that is going to foul you up.

Incidentally, I have found companies to be very uncooperative about telling how they make compounds. If they would only tell us this is one way of guessing what contaminants might be present. Has anyone else had this experience?

SANCHEZ: Many companies don't even know where they get their materials. If you ask Sigma (Sigma Chemical Co., St. Louis, Mo.), and ask them about their compound, they have gotten it from somebody else. They don't know themselves how the others made it.

ORGEL: I think a more severe problem of the same type concerns the purchase of radioactive materials, because often they are less than 50 per cent pure.

USHER: There has been recent correspondence on that in the literature, hasn't there?

ORGEL: Let us have some views (which perhaps shouldn't be published for reasons of libel) concerning companies which are less reliable in their supply of radioactive compounds if there are such companies.

What has been the experience with labeled nucleosides and nucleotides?

LEMMON: We have all had our bad experiences. If you combine gas with paper or thin layer chromatography, both volatile radioimpurities and nonvolatile radioimpurities can be detected. We must all chromatograph commercial labeled compounds before using them. The company may have produced a perfectly good labeled product which they or you leave on your shelf for some months. This leads to self-radiolysis. All these companies have to contend with this problem.

PONNAMPERUMA: This is especially true with the amino acids. The Radiochemical Center at Amersham, for example, sends material frozen. Any labeled amino acid will give five or six different spots on autoradiogram. Self-radiolysis is a serious problem.

ORGEL: We probably have spent enough time discussing nucleotides. Could someone summarize roughly our thoughts about the relative strengths of techniques? What constitutes an ideal identification? What constitutes a minimal identification?

USHER: Would someone sum up how we may get more volatiles? What are the best volatile nucleoside derivatives?

I'm not at all sure. I have worked in phosphates for many years, but I still don't know about volatilizing nucleotides, the stability of various derivatives, nor what are the preferred volatilization methods.

PONNAMPERUMA: A technique we have tried is to make the trimethylsilyl derivatives of the sugar. It is not ideal, but so far these are best. You make substitutions in a number of places.

ORÓ: One of the best separation methods, in the future, may be based on high pressure liquid chromatography which has comparable capability to gas chromatography. I'm referring to capillary columns that have a stationary phase, either solid or semi-solid, and can perform the separation of compounds by means of a liquid solvent, rather than by means of a carrier gas. The same people that do research in gas chromatography have been involved in the development of this method. I hope the next five years will bring developments in this area, so that phosphorylated derivatives or compounds with molecular weights too high to make volatile derivatives can be separated by this method. They can be detected at the end by universal or specific detectors.

With this new technique it should be possible to characterize peptides of more than 10 amino acids and nucleotide polymers with a large number of mononucleotides. For peptides the upper limit by gas chromatography today is around 10 amino acids. Rather than try to make volatile derivatives, I think we'll get sophistication in the liquid chromatographic method that will permit characterization as good as we are doing today by combined gas chromatography-mass spectrometry.

USHER: This is like Sandy Lipsky, using high pressures and glass beads coated with ion exchange resin. For short sequences there is quite a bit of encouraging work from Cantor and others (Cantor and Tinoco, 1967). One can determine sequences of trimers, based on differences in optical properties such as circular dichroism and optical density. Even with the same three nucleotides, if the sequence differs, so do the properties.

ORGEL: Is this not limited to the standard 3'-5' linkages?

USHER: Yes, I think so.

ORÓ: Another new method is fluorine NMR. I think Dr. Bayer's group has determined sequences up to 10 amino acids by signals from peptides which have been appropriately derivatized. The sequence of amino acids can be determined without destroying the compounds.

ORGEL: Joseph, since you started this discussion, would you please summarize your views on good thorough nucleotide identifications as well as acceptable ones?

NAGYVARY: The UV spectrum is cheap and fairly characteristic. An NMR spectrum, in my view, is an absolute necessity. Together with easily measured chromatographic mobilities these two techniques provide a high degree of accuracy.

If the alpha-beta configuration is to be definitively distinguished we can make a chemical derivative of nucleoside which can be characterized with a few milligrams compound. These techniques actually characterize the natural nucleosides. You may do five more, but I feel that UV, NMR, chromatographic mobilities and paper electrophoresis ion exchange resin, completely characterize the natural nucleotides or nucleosides.

ORGEL: I would have added the enzymatic techniques, which, I think, are very powerful and in many cases very easy.

PONNAMPERUMA: The stability of various configurations to acid or base hydrolysis is also very good to look at.

NAGYVARY: This certainly can be done, particularly if you don't have an NMR. Of course, NMR easily distinguishes those.

USHER: I think C^{13} NMR and the use of these paramagnetic chelates to spread out the proton resonances will eventually be very powerful analytical techniques.

Porphyrins

ORGEL: Thank you. Two topics are left for discussion. One concerns other molecules – porphyrins, and so on – or perhaps you have had enough detailed chemistry, and would rather turn to more philosophical matters.

Shall we go on to porphyrins and identification of other molecular classes or rather (along the lines that Carl suggested) shall we try to place quantitative estimates on the power of these analytical techniques?

NAGYVARY: I believe a general discussion is more valuable because many times we don't know what compounds we are going to get. We should not expect only the occurrence of natural products in prebiotic experiments. Some compounds which are not now natural products may have had an important role 3 billion years ago. This must be characterized.

ORGEL: This is a third topic: namely, the identification of unknowns, rather than the confirmation of an assigned formula. That takes us into all sorts of horrible areas of structural organic chemistry.

LEMMON: Before we leave this, you might ask if anyone has any identification method that's particularly good for any compounds that haven't been discussed yet.

ORGEL: Perhaps someone would discuss fluorescent methods, for example. Some things fascinate me personally. I mustn't force them on you, but I would like to hear what people feel about porphyrin identification.

PONNAMPERUMA: Unfortunately, two experimenters on porphyrin analysis are not here.

ORÓ: My estimation is that most of the work done has been with relative methods.

MARGULIS: Porphyrins have not been made under prebiotic conditions, is that correct?

PONNAMPERUMA: The identifications of porphyrin in prebiotic

systems have been done based on gel separation, fluorescence spectroscopy, and demethylation with sulfonic acid (Hodgson and Ponnamperuma, 1968). Identification would be on far more solid ground if we were able to isolate enough material from a prebiotic system for mass spectrometry.

The identification of porphyrins in lunar samples is only very preliminary. Gordon Hodgson himself would present it that way. On the basis of what he has found there is a spectrum suggestive of prophyrins (Hodgson, et al., 1970).

In both cases there hasn't been enough material to use any other technique. The circular dichroism technique is sensitive and is available now, but I think at least 100 nanograms of material are needed. The identifications presented by Hodgson have been based on about 10 nanograms.

ORGEL: Then we don't really have a satisfactory technique for small quantities of porphyrins.

PONNAMPERUMA: Not completely.

BARGHOORN: Cyril, on theoretical grounds, what is wrong with the methyl sulfonic acid technique?

PONNAMPERUMA: Nothing. It's a useful technique. A metal ion is removed giving a variation in the spectrum characteristic of a porphyrin. But this doesn't really satisfy the chemist. We have no structural identification.

In the case of the lunar sample we asked NASA for 50 more grams to confirm the identification on the Apollo 12. I suppose they haven't given it to us because NASA feels rightly ten years from now there will be a much better technique. At the moment they don't want to lose those 50 gm.

YOUNG: Isn't the porphyrin situation really even worse than that? Now both Hodgson and Rowe find prophyrins in each other's blanks. I think total chaos is a fair statement!

ORÓ: Rowe told me about some scattering phenomena overlooked in the past which provide an answer very similar to the answers obtained so far.

YOUNG: It's not that straightforward.

SCHOPF: At the Lunar Science Conference that explanation had not clearly been established by him.

PONNAMPERUMA: Since the Lunar Science Conference of Apollo 12 as an exercise in preparation for Apollo 14 we had a sample of the Onverwacht shale prepared in our laboratory. A sample was given to Hodgson and another to John Rowe. Both found porphyrins in the sample by each other's technique.

YOUNG: It has gone one step further.

PONNAMPERUMA: The next step was when John Rowe used less

material 1/10 the concentration – Rowe couldn't detect porphyrins, but Hodgson did.

YOUNG: That's another complication.

PONNAMPERUMA: I don't know about it.

YOUNG: I just reported it the day before yesterday!

Rowe made up a sample in his laboratory which he has given to Hodgson. They analyzed each other's samples, and in the case you mention they got comparable data, except Hodgson finds porphyrin in Rowe's blank and Rowe finds porphyrins in Hodgson's blank. In the material Rowe made up for Hodgson to analyze, they disagree by two orders of magnitude in quantitation. They are not even close right now.

ORGEL: Apparently the porphyrin field is completely open – any particular information can either be believed or not.

YOUNG: How do we resolve this? It's a mess right now. We're trying to get a third investigation.

SCHOPF: To have three different sets of answers. Leslie, in answer to your question about the direction for the conversation to go, I would like Bob Sanchez's numbers.

Relative Value of Analytical Methods

ORGEL: Bob, you have an invitation.

SANCHEZ: This is just a row of numbers, almost arbitrarily chosen. There are many problems. The value of an analytical method depends on what you are analyzing for, of course every analytical method has an extreme range in reliability. This list can only be on the basis of my experience, the techniques I use and the way I use them. (Table 4)

Here is the name of a technique, and a number I think generally expresses the reliability of that technique for the identification.

TABLE 4. ESTIMATED RELIABILITIES OF VARIOUS TECHNIQUES
FOR THE IDENTIFICATION OF SINGLE AMINO ACIDS
(BY COMPARISON WITH AUTHENTIC STANDARDS)

X-ray diffraction	0.99
Enzymatic analysis	0.95
High resolution mass spectrometry	0.95
Combined gas chromatography-mass spectrometry	0.95
Low resolution mass spectrometry	0.9
Infrared spectrophotometry	0.9
Mixed melting point	0.9
Nuclear magnetic resonance	0.8
Automatic amino acid analysis	0.8
Gas chromatography	0.7
Paper chromatography	0.5
Melting point	0.3

ORGEL: And these are based on intuition?

SANCHEZ: More intuition than anything else; my own experience with these techniques.

With an individual amino acid in some small amount, subjected to these various methods, I would assign these figures, which I have made pessimistic. They are on the low side.

ORÓ: Why don't you leave out melting point? I don't think it has much value if mixed melting points aren't done.

SANCHEZ: The value is very low.

ORGEL: Are there any negative values: techniques that point in the wrong direction?

SANCHEZ: My numbers don't mean much, but what's interesting is whether such a system can be developed.

WOLMAN: Do you mean the mixed melting point of amino acids or derivatives of amino acids?

SANCHEZ: That means melting point of an isolated amino acid or a derivative in a mixture with an authentic sample.

WOLMAN: Then I think you are very, very optimistic in the mixed melting points.

SANCHEZ: Should the values be lower?

WOLMAN: It should be much lower, because most amino acids, if not all, melt at about 200-250°C, and —

LEMMON: They hardly melt. They decompose.

WOLMAN: Yes, they really decompose. I would put zero on both melting point and mixed melting point.

MILLER: Let's change its meaning to derivative in the case of amino acids.

PONNAMPERUMA: In general, you can prepare a low melting derivative.

WOLMAN: Write melting point of the derivative then.

ORGEL: What is the reliability value for a survey of the literature? [Laughter]

SAGAN: Can we all agree on such numbers? If so, they would be used in the following way.

Assume all techniques are independent. On an automatic amino acid analysis, you might have a probability of being wrong of 0.2. A gas chromatographic analysis might have a probability of being wrong of 0.3. A paper chromatographic analysis on the same sample might give a probability of being wrong of 0.5. The joint probability of being wrong with all three techniques would be (0.2 x 0.3 x 0.5) or 0.03: a 97 per cent probability of being correct.

USHER: But you don't know the cross-correlations here.

SAGAN: As I say, assuming statistical independence.

ORGEL: But suppose Bob had put up all those numbers with one extra zero in front of them.

SAGAN: Of course it wouldn't change the values, if it were on the left-hand side of the decimal!

ORGEL: Everything could be halved and –

SAGAN: Yes, giving you 98 or 96, or something, instead of 97 per cent.

ORGEL: The difference between 99.9 per cent reliability and 99.7 is significant, but the difference between 50 per cent reliability and 48.5 is insignificant.

SANCHEZ: These figures can never be fixed, they are a function of the technique.

SAGAN: But it would be useful to know the combination of three independently low weight techniques can give you a reasonably high –

ORGEL: I don't think I understand what you are saying, Carl, because everyone has always realized that to use more than one technique is better than to use one. What are we doing more than that?

SAGAN: Each laboratory has a certain array of analytical instrumentation available to it. Not everyone has a GC-MS. The question is: Is it possible to have a combination of inexpensive or more readily available techniques, which give you the same reliability for an identification as GC-MS?

ORGEL: This discussion is too philosophical. People are interested in various things, but they are more interested in the chance of being wrong in particular examples. These general numbers are not valid at all. This merely tries to make quantitative things which everyone knows intuitively.

SAGAN: Speaking for myself, I do not know them intuitively, and, therefore I find this approach useful.

ORGEL: If it were really possible to decide on valid numbers, of course this would be much better than intuition, but I think the field is such that you cannot use a single number to characterize the technique over a wide range of compounds.

SAGAN: But Bob has gotten these numbers from somewhere. They must represent some impression –

ORGEL: His experience, probably his experience with nucleotides.

SANCHEZ: No, no. I'm trying to base this on amino acids. Let us pick some particular amino acid, say, phenylalanine, that we have had experience with and see if these figures make sense.

LEMMON: Are you only applying these to amino acids?

SANCHEZ: Yes.

ORGEL: What do you mean by "enzymatic analysis"? Which enzyme?

PONNAMPERUMA: Peptides, obviously.

SANCHEZ: This was brought up before (p. 31). Such techniques involve adding the unknown amino acid to a system deficient in that particular amino acid —

MILLER: Has this really been done?

SANCHEZ: Isn't this done in the resolution experiments? A synthetic amino acid will be D-L mixture, and, if this is metabolically utilized — exactly half of it will be used, giving you an optically active D-amino acid left over.

BADA: Is this necessarily true? If you are going to talk about determining D and L with the enantiomers. I have to rate enzyme analysis near the absolute bottom. I and others have tried to use D-amino acid oxidase and find I get 100 times more amino acid out of the system than I put in, just because the enzyme is loaded with amino acids.

SANCHEZ: I have never done the technique myself.

PONNAMPERUMA: This is exactly the point I raised earlier: are enzymes free of amino acids required?

ORÓ: May I attempt to explain the confusion? Bob was referring mainly to microbiological testing methods. These are standard, used for many years. Say you want to determine phenylalanine in a mixture as complex as you wish. If you have a mutant bacterium which is dependent on phenylalanine, you add it. This allows you to determine whether it's phenylalanine, and how much.

ORGEL: That's a different technique, and it is a good one. It was used on the distinction between alpha and beta cytidine, and it's a very good technique. But it is different from the technique of using a purified enzyme.

MILLER: You put the reliability of x-ray at 0.99. Is that an x-ray powder pattern, or —?

USHER: Single crystals.

ORGEL: To determine realistically these values, we have to include times. Against that 0.99 for x-ray crystallography, if you haven't got the machine, the time is infinite. Even if you have got it, you still probably need three months for an analysis whereas to do the enzyme technique you probably need an afternoon.

USHER: If you get a crystal of the substance and just want to determine if it's phenylalanine, you can take a single crystal and get the diffraction pattern and see if it's right.

ORGEL: How long would it take to set the crystals up accurately? A couple of days?

SIEVER: Ordinarily, mounting the crystals and getting a diffraction pattern can be done very quickly.

ORGEL: Then this would be a good technique?

MILLER: But it would be enormously expensive.

USHER: You only need a millimeter crystal.

ORGEL: And you can only do it if you have a friend with an x-ray diffractometer next door.

SAGAN: The cost and the time needed for each technique is one question, but I wonder: Does everyone agree to the reliability of Bob's numbers.

PONNAMPERUMA: No! For some amino acids, those that don't separate out on the automatic amino acid analyzer, this technique would be way, way down.

MILLER: If you see something in an amino acid analyzer with characteristic elution time, of the 20 protein amino acids the identification is 10-20 per cent. That's my experience. For example, when your unknown appears at the serine or isoleucine position on the amino acid analyzer, you can be pretty sure it's not serine or isoleucine.

ORGEL: My impression is that if these numbers were right, prebiotic chemistry would be a lot easier than it really is.

LEMMON: They are really very optimistic numbers, not pessimistic.

NAGYVARY: Is it not true that in protein synthesis there is a very specific enzyme for every amino acid?

MILLER: Isoleucine and valine will be confused if you are not very careful.

NAGYVARY: But these two are probably easily separated on the amino acid analyzer, or gas chromatography.

MILLER: Yes, but there are other confusions. There is a whole literature on compounds that will substitute for various amino acids in amino acid activating systems. Am I not right?

ORGEL: Yes. You see, we must distinguish the problem faced by the biologist, who has only the 20 naturally occurring amino acids — and that of the prebiotic chemist, who has not only those 20 amino acids, but an infinite number of others. Biological systems can usually take care of 20 amino acids, and know very well to distinguish between any pair of the 20. Such systems are often not at all good at distinguishing between one of the 20 and something from outside the set. Therefore you have to be very cautious in applying microbiological methods — and even more so, enzyme methods — to substances which are not from biological sources.

The literature claims that adenosine deaminase is specific for adenosine, meaning it doesn't work on cytosine! That is very different from saying that it works for adenosine, and adenine can't be replaced by another base, or ribose by another sugar. We have one example when alpha-adenosine is used instead of beta-adenosine, it does react, but about 50 times slower. Isn't that right?

SANCHEZ: I'm not sure.

ORGEL: I think it did deaminate, but much more slowly.

SHELESNYAK: Are there a number of other methods to add to this list? On the basis of visceral reactions and experience, can we get a range of probabilities from the group?

SAGAN: That would be very useful.

ORÓ: Leslie is saying we don't have the information necessary to do a job that I would put my signature to. It's fine to do something like this on a tentative basis, but for each compound, each situation, there is a different set of numbers.

Three methods may have a synergistic effect in terms of the validity of the final result, or not, or the reverse. They may not be independent. I think the complexity of the situation suggests we wait ten years to do what you want, Carl. If people are very mindful of the situation, enough experience may be accumulated to eventually give you the numbers you want.

SAGAN: I agree that ultimately we need a chart for every compound that one might conceivably be interested in identifying. That also will take more time than the ten minutes we have before lunch today.

But people ought to put down the ranges they think correct and it may turn out that the ranges are sufficiently constrained that one learns something.

For example, isn't 0.95 for high resolution mass spectrometry overly pessimistic? Wouldn't it be significantly better than that?

USHER: But, Carl, these numbers will range from what they are (Table 4) down to zero, depending on the situation.

SANCHEZ: Can you imagine a circumstance where x-ray diffraction would give you a figure less than, say, 0.9?

USHER: Yes, if the molecule is big enough.

SANCHEZ: We're talking about amino acids.

USHER: Okay for amino acids.

ORGEL: If it's polymorphic, and you crystallize it on two different days, you might get two different crystals formed, in totally different patterns.

SANCHEZ: We can't use that technique unless we have a single crystal.

USHER: But we don't say that. We take the new pattern and say: we don't know what this is. Let's solve it. Then you discover it's a different crystal form of the same compound.

ORGEL: You are suggesting solving the crystal structure, which takes three months.

USHER: If we have to.

ORGEL: That doesn't seem to be a practical method of doing this sort of work.

USHER: But we usually do not have to do that, because the polymorphic forms are known from the literature. That 0.99 could be very wrong. For instance sometimes you can't tell the difference between a thiol and an oxygen, especially if it's in a crystal form. You may have an alternation of thiol and oxygen, and not know it. Even x-ray analysis can give you false results.

SHELESNYAK: Is this also true for low temperature x-ray analysis?

USHER: Yes. You can usually tell a more electron dense atom, which means it is thiol instead of oxygen; if there is an alternation — in one place oxygen and in another sulfur — you may have trouble.

MARGULIS: In your example all that is needed is elemental analysis. The problem is: How often do you have a millimeter size pure crystal of any relevant "prebiotically produced" material?

Those of us interested in what happened on the primitive earth have this to face. What do these experiments yield? Under what conditions do you get pure materials that can be handled crystallographically? Porphyrins seem to be negative, and sulfur amino acids on the border line. What prebiotic materials have we really made in quantity?

BARGHOORN: There are alkane hydrocarbons.

USHER: All the values (Table 4, Sanchez) can go to zero, depending on the situation. It's meaningless to discuss it.

ORGEL: Anyone who thinks this discussion is meaningful, let's continue.

IDENTIFICATION OF ORGANIC MATTER IN NATURAL SAMPLES

SIEVER: May I suggest that we not just discuss prebiotic synthesis, but in general the problems in sampling. We are already discussing porphy-rins in lunar samples. Apollo has provided at least one example of very care-fully collected materials.

John Oró and I have talked about problems of contamination of natural samples — and this issue is by no means clear.

A most important issue involves the theory that rocks are impermeable. If the onion skin on the outside is peeled off you will eventually reach a part on the inside which is immaculate and therefore represents the original con-ception!

I doubt this; Elso Barghoorn and Bill Schopf and others probably have information on this.

Some may have ideas too on working an inverse spiking method, rather than trying to get the completely uncontaminated center. I mean some way in which you simply compare what the contaminants are, and then start sub-tracting away to find out what the unnatural materials finally might be.

BARGHOORN: Are you talking about geological contamination?

SIEVER: Any material from the earth.

ORGEL: What minerals and materials have you in mind? Are you thinking about hunks of Precambrian rock?

SIEVER: I'm talking about iron-nickel chondrites, carbonaceous chrondrites*, the Precambrian rocks, lunar samples. This problem applies to any of these; most particularly on the earth, there is a possibility now of contamination in every old rock.

ORGEL: Let's stay with one topic.

SIEVER: Then lets discuss the cherts.

ORGEL: Will a geologist tell us about contamination by water seeping through rocks and so forth?

BARGHOORN: The porosity, the intrinsic permeability as seen in scan-ning electron microscope pictures of chert, is quite evident. Cherts obviously do not have a highly fused structure but are a porous system at that level of observation. Data, kindly made available to me by Robert Clarke of the Mo-bil Research and Development Corporation, Dallas, Texas, are shown in Table 5.

---

*An extensive discussion of the microspheres of the Orgeuil carbonaceous chondrite was led by Professor Barghoorn that evening, which unfortunately was not recorded. The evi-dence that the microstructures are abiotic organic particles produced within the meteorite and uncontaminated by terrestrial life forms has been summarized in a paper containing over 100 references to this controversial problem. See Rossignol-Strick and Barghoorn, 1971.

TABLE 5. POROSITY AND PERMEABILITY OF ANCIENT "HARD"
AND "IMPERMEABLE" ROCKS.

Source of Rock*	Porosity** %	Density, Bulk	gm/cc Solid	Gas Permeability (md)
Fig Tree Formation	0.36	2.64	2.65	<0.1
Bitter Springs chert	1.5	2.65	2.68	–
Chert from Kewatin rocks	0.77	2.69	2.71	–
Gunflint chert	0.34	2.64	2.65	<0.1
Maravillas chert	1.2	2.54	2.57	0.18
Algoman granite	0.91	2.63	2.66	<0.1
Biwabik Iron Formation	2.4	3.11	3.18	0.1

*All of above are Precambrian except for the Maravillas chert, which is Ordovician.

**Porosity measured on Kobe Porosimeter.

Granites are quite permeable. The Algoman granite, one of the lightest granites available, has a much higher permeability than the Gunflint chert; but they all show permeability — what did Nagy find with the Onverwacht cherts? One volume replacement in a billion years, as I recall. With hydroxylated compounds which can tumble into these minute porosities, the whole amino acid problem in rocks is already on swampy ground. I would argue that the alkane hydrocarbons are a different matter.

SCHOPF: Several workers have done analyses of porosity in a variety of ancient sediments. Such data are in the literature (e.g., Smith, Schopf and Kaplan, 1970), and, in fact, all rocks are permeable to some degree. With an essentially infinite or very long period of time, there is an opportunity for contamination to occur.

Let me summarize some data, based on analysis of one rock type from one locality by three different laboratories, which suggest the magnitude of the problem. The rock specimens were from the Gunflint chert, a middle Precambrian deposit that Elso Barghoorn has worked on for years, which contains a really remarkable assemblage of microorganisms about 1.9 billion years in age (Barghoorn and Tyler, 1965).

One group (Smith, Schopf and Kaplan, 1970) found about 0.029 to 0.039 parts per million of total alkanes, including surface contaminants. A second group (Oró, Nooner, Zlatkis, Wikstrom and Barghoorn, 1965) found 5 parts per million, and a third group (Van Hoeven, Maxwell and Calvin, 1969) observed 10 to 100 parts per million. Oró et al. (1965) reported no evidence of odd carbon preference among the alkanes; Van Hoeven et al. (1969) reported alkanes with an odd number of carbon atoms in the C_{23} to C_{29} range; and Smith et al. (1970) reported odd carbon preference in the C_{27} to C_{33} range. These data were all obtained on rocks from the same locality.

ORGEL: Does "the same locality" mean very close together?

SCHOPF: Yes.

Two were collected by Elso; one was collected by me, and they were all from the same outcrop, perhaps 100 m^2 in area.

MARGULIS: Are the microflora essentially identical in the rock samples?

SCHOPF: Apparently so. Incidentally, it seems worth noting that there was a difference among the three groups in the amounts of extractable organics attributable to surface contamination. One group (Van Hoeven *et al.*, 1969) apparently didn't concern themselves with this problem. A second group (Oró *et al.*, 1965) did a benzene-methanol extract and found that less than 1 per cent of the total hydrocarbons recovered from the chert could reasonably be attributed to surface contamination. The third group (Smith *et al.*, 1970) reported values of 82 to 89 per cent that they could get out by exhaustive extraction with a benzene-methanol solvent solution.

Who is right and who is wrong doesn't really matter. Here were three experienced groups with rather similar procedures that came up with highly different answers.

I conclude there is considerable variability of *in situ* contamination. I don't think any group was particularly sloppy in the laboratory. Everybody knew what he was doing, but presumably there is considerable variability. You are confronted with this problem which goes back to this permeability business.

It is a mess.

BARGHOORN: I think this problem is insoluble. First there are these microporosities in cherts, which go down to hundreds of Angstroms. Variability depends on your level of discrimination. In one field of the microscope there are highly different amounts of organic material. Even in a 30-micron-thick section the organic material per unit rock volume is highly variable. These samples would have to be homogenized. Taking them from 2 yards apart doesn't mean much. Total homogenization is required to make them comparable.

ORGEL: Have I missed the point? What is your conclusion?

BARGHOORN: My conclusion is *a priori* and didactic and dogmatic: You cannot depend on any of these results from rocks.

ORGEL: Does this apply to hydrocarbon analyses?

BARGHOORN: The alkane — any lipid count — is much less suspect.

ORGEL: Then the alkanes are most reliable, and everything else is hopeless?

BARGHOORN: Equivocal.

SIEVER: Perhaps Jeff will talk about the racemic mixtures. Racemic mixtures in old rocks can be used as pretty good evidence of the fact that they may have been original.

BARGHOORN: But this doesn't put a limit on the age of contamination.

SIEVER: I'm talking about Precambrian. Is there any amino acid that won't racemize in the time elapsed since the Precambrian?

SCHOPF: Glycine!

BADA: I think it's obvious that the amino acids in the Precambrian cherts are not indigenous. The latest data indicates the amino acids isolated from the Gunflint or the Fig Tree cherts are 90 to 98 per cent the L form. From my work on racemization kinetics in the geological systems, these amino acids should be completely racemic after times no longer than about 25 million years. This strongly suggests these amino acids are secondary.

ORGEL: Does the rate of racemization depend on conditions such as humidity? Do racemization rates applying to one set of conditions apply in a rock?

BADA: I plan to talk about this tomorrow.

ORGEL: Before we can accept what you say, we need to be convinced but let's leave this until tomorrow.

BADA: Fine.

Both Cyril and John Oró have been doing work on meteorites (Cyril can tell us about that) where definite D-L mixtures are found — isn't this probably a pretty good indication that contamination is not a problem?

MARGULIS: Is the discrepancy due to recent laboratory contamination or geological contamination?

BADA: A combination of both, I think.

SCHOPF: And it's very difficult to determine the primary component. I think most is due to *in situ* contamination. Almost all workers have reported about the same spectrum of amino acids in extracts of the very ancient rocks that have been studied; it looks as though the amino acids are in the rocks, although they are not as old. Probably the amino acids are very much younger than the rocks.

PONNAMPERUMA: In your own work didn't you find L-amino acids, and suggest they probably came in at a later date?

SCHOPF: Yes, unless amino acids are geochemically stabilized in the L-configuration in some as yet undefined way, the occurrence of just one isomer — the L-isomer — seems highly suggestive of a relatively recent origin. I don't see a ready source for recent racemic contamination; that is, a D-L mixture of contaminating amino acids. That, of course, is the value of the Murchison work (Koenvolden *et al.*, 1970).

PONNAMPERUMA: I'm not talking about Murchison. I'm talking about the Fig Tree and the Gunflint work where the amino acids are largely L, suggesting they were recent, less than 25 million years old, from what Jeff Bada said, right?

ORÓ: I think the L-amino acids both in Precambrian sediments and in meteorites are the products of hydrolysis by microorganisms. Several of our meteorites were further extracted with solvent, then with water, and then hydrolyzed, which gave us predominantly L; I think that hoping to remove contamination by water extraction and solvent extraction, we are leaving there what is the solid contamination by residues of recent microorganisms, which still retain essentially all the L configurations.

SCHOPF: Why doesn't anybody find diaminopimelic acid, which occurs in bacterial cell walls and can be identified rather easily.

MILLER: But there isn't only one kind of wall, the gram positive bacteria differ from the gram negatives.

MARGULIS: Most any bacterial sample should have it.

SCHOPF: Most of the shales have diaminopimelic acid, yet no one has reported it in the ancient rocks.

ORÓ: We have not looked for this.

SCHOPF: Well, I have looked for it.

BADA: How?

SCHOPF: I ran standards with an amino acid analyzer, also with GC, making derivatives and doing gas chromatography, and I did not see diaminopimelic acid.

BADA: Doesn't it come off on the analyzer right about where alloisoleucine does?

SCHOPF: I'm not sure, but I remember I convinced myself and an assistant that we would have seen it, but we didn't, which I thought was strange.

BADA: We have found that unless you use a 150 cm column or a greatly modified buffer scheme, your diaminopimelic acid comes up right smack on top of alloisoleucine.

ORÓ: So we may have it!

ORGEL: That is very interesting, obviously a nice way of distinguishing between several different situations.

SCHOPF: It is also a very nonstable amino acid; it would not be expected to stick around in sediments.

MARGULIS: It's also going to be in recent bacteria in quantity.

SCHOPF: I want to find out something from Elso before lunch.

ORGEL: We're going to go off for lunch in about three minutes.

SCHOPF: Elso, why do you have faith in hydrocarbons versus the nonhydrocarbons? The data that I read off were entirely devoted to n-alkanes. I was wondering how one can reasonably distinguish between those two categories? I'd like to, but I don't understand the logic.

BARGHOORN: I think it's a matter of poly versus nonpoly compounds. I'm not an organic chemist, but I think any hydroxylated compound is much more apt to be suspect because of the tumbling effect into microporosities.

ORGEL: Is that an experimental fact?

BARGHOORN: No, a suggestion.

MILLER: Is there evidence that hydrocarbons diffuse, or go through porous materials more slowly, than the hydroxylated compounds?

BARGHOORN: We don't have a petroleum chemist here, but I think that the diffusion rate of alkanes into rocks would be infinitely less than the diffusion of hydroxylated compounds.

SCHOPF: You have an infinite amount of time.

BARGHOORN: That's another problem.

The fine structure of rocks under the scanning electron microscope is in a state of infancy.

But leaving that aside, the permeability-porosity measurements are really quite alarming in terms of the intrinsic capacity of diffusion through rocks of any category. I'm not talking about clays.

SIEVER: Calculations using any kind of reasonable diffusion coefficients for what happens in two billion years would wipe out the difference between alkane and non-alkane hydrocarbons.

PONNAMPERUMA: Both Elso and Bill have found microfossils in these rocks; here is one point we can be sure about.

BARGHOORN: We know that there is organic stuff here in great quantities.

PONNAMPERUMA: Are you looking at the composition of the micro-structures?

SCHOPF: In the Precambrian materials we have a small amount of organic matter, and the indigenous syngenetic problem is fierce. The only reliable method will be on the chemistry of organic material that must have been deposited syngenetically; i.e. chemistry of the microfossils themselves. This is difficult but I have been trying it recently anyway. I still worry about organic matter entering the rock, and combining with the microfossils at some later time. This sort of problem has really not been attacked before.

ORGEL: Thanks. We must continue this discussion after lunch.

ENERGY SOURCES AND THE PROBABILITY
THAT LIFE COULD HAVE ORIGINATED
Wednesday Afternoon Session

Dr. Stanley Miller opened the afternoon session.

MILLER: Our Chairman this afternoon is Carl Sagan.

SAGAN: Thank you, Stanley. This session is called "Sources of Energy", but our main focus will be on Dr. Hulett's article (Hulett, 1969) which was mailed to all of us, and which in some quarters has been described as demonstrating that the origin of life couldn't have occurred. I'm sure this was not his intent. Instead, it was to call attention to with potential problems in the understanding of the origin of life.

Le Compte de Nouy's *Human Destiny* is one of the earliest books on the origin of life (although not at all up to contemporary standards). In it he argued essentially as follows for the existence of God: In each position of any protein there's one chance in twenty of a given amino acid being there randomly. If there are a hundred peptide linkages, the chance of such a molecule having been spontaneously assembled is, therefore, 20^{-100} power. This is a smaller number than the reciprocal of the total number of elementary particles in the universe! Since no protein could have been spontaneously assembled, protein formation had to be guided, and therefore God exists. No mention was made of natural selection as a probability sieve. Although this is not the tenor of Dr. Hulett's arguments, there is a thin edge in the demonstration that I can't resist drawing attention to.

The question of energy sources is a very important one. Some investigators use particular energy sources because they are experimentally convenient, rather than because they are those most likely to have been important on the primitive earth. It's important to concentrate on just what is the most geophysically plausible scenario for energy sources on the early earth. There are two quantities to be considered concerning energy sources. One is the amount of useful energy available for prebiological organic synthesis. The other is the efficiency of the synthetic process using that energy. When energy sources are compared the product of energy times efficiency, not just the energy, not just efficiency, must be compared.

May I call on Dr. Hulett to discuss this further: Dr. Hulett?

HULETT: I hope that the fact that Carl and I are on opposite sides of the table is just a statistical probability, and not a design! I'm not sure whether my role here is Discussion Leader or Clay Pigeon because I'm sure you're all going to take potshots at me!

My position is quite different than most of you, in the sense that I have not done any experimental work whatsoever on the origin of life but for me it is just sort of an avocation. I'm probably here because some of you have good press agents. My daughter showed me a statement in her high school biology textbook with which I disagreed. It was similar to the quote by G. G. Simpson (Hulett, 1969): "Virtually all biochemists agree that life arose spontaneously from nonliving matter, and would almost inevitably arise on sufficiently similar planets elsewhere." The word "inevitable" just didn't sit right with me, and after I looked into it more, I decided to write up my conclusions and see if somebody was willing to publish it. They did, and here I am.

Before you start in on the article, I'd like to start in myself. Please let me first indicate where changes are certainly needed and then open myself to questions. We can have a general round table discussion or I can discuss energy sources I think were probably most useful. First, on page 58 —

SAGAN: Does everyone have a copy of the article? Might you just briefly outline the arguments you are changing?

HULETT: This is not an argument, but just a figure (something Carl and I have discussed before): the energy available for meteorites or for meteoroid material. First, on page 58 the word should not be "meteorite" but "meteoroid material" or "meteoroid".

The energy available is probably greater than what I have said, by maybe an order of magnitude.

SAGAN: Because what if the efficiency is much higher from such a process?

PONNAMPERUMA: What is the latest figure available?

SAGAN: I agree that it's about an order of magnitude larger than his figure.

PONNAMPERUMA: Only an order of magnitude?

SAGAN: We are really guessing what it was 4 billion years ago, but I think that it is not being too pessimistic if this number is off by a factor of 10.

MILLER: The figure he has is 4×10^{-5} watts per square centimeter per year. What will you raise it to — ?

SAGAN: 10^{-3} is a good number.

MILLER: About the same as cosmic rays, in other words.

HULETT: It is small compared to cosmic rays. On page 59 I say solar ultraviolet radiation is generally considered the most likely energy source for prebiological synthesis. Some people here will disagree with the word (generally," I probably should say "UV is considered by many", or something. It's a minor point.

SOFFEN: It's like the word "inevitable."

MILLER: In a sense it's more than a minor point.

HULETT: As far as the paper is concerned it's a minor point. We can discuss energy sources this afternoon, but as far as the paper is concerned it is not important.

On page 60 I neglected the formation of formaldehyde from hydrogen and carbon monoxide. I have no excuse, since I had in the Groth (1937) reference which I quote. I had known that he had a higher quantum efficiency for that particular reaction than for methane and carbon monoxide — about 0.5 instead of 0.1.

Philip Abelson and Thomas Hoering in some electric discharge experiments made an attempt to provide UV radiation — I don't know how good their attempt was, but there should have been some UV radiation there. They found no formaldehyde synthesis from atmospheres containing carbon monoxide and hydrogen. I should mention something not in my paper: formaldehyde, in addition to the absorption bands that Carl was talking about yesterday, has very strong absorption bands further down in the ultraviolet, with extinction coefficients about 1750-1800 Angstroms.

A change of an order of magnitude in the formaldehyde concentration won't make much difference anyway. If however there are production rates on the surface that Jerry Hubbard told us about last night, we have a different ball game. We have to decide how much is absorbed and how much released; how will he be able to calculate that?

Also, on page 60, paragraph 3 the word "altitude" was transposed to "latitude"; again a minor point.

On page 61 I discuss an attempt to make an estimate of the highest possible HCN production, instead of "highest possible" I should say "an optimistic prediction."

On page 62 I discussed phosphate incorporation. Certainly now the recent work of Lohrmann and Orgel should be mentioned. Taking into consideration their work I would change my last sentence from "All origin of life experiments in which organic phosphates have been generated have used either condensed phosphates or extremely high concentrations of soluble phosphates" to "Most" instead of "All". I do think theirs is the most promising experiment I have seen so far.

On page 64, in the discussion of electrical discharges, I should have included the work that Cyril and Ruler did on the synthesis of amino nitriles. I need now also to include the work that Carl, Bar Nun, and Bauer have done on shock synthesis. Shock waves in these latter experiments were amazingly efficient, as they indicated in their paper (Bar-Nun *et al.* 1970). If the efficiency is actually that high, shock synthesis might even be used as a commercial source of amino acids.

Certainly, this work should be repeated and extended. I personally would like to see it repeated with a closer simulation to the initial components of the primitive atmosphere. (Surely Carl will say something about that later).

PONNAMPERUMA: How did you miss the amino nitrile paper?

HULETT: I don't know.

LEMMON: Cyril, you didn't send him a copy!

HULETT: It wasn't published in anything I usually read.

In the middle of page 70 I said the life of HCN is essentially infinite at low temperatures. I should probably have said it is "very long at low temperatures."

LEMMON: How low are the temperatures?

HULETT: That's just the trouble. I don't know. Now you have my corrections before you start on yours.

I really meant to draw attention to a number of areas where I feel more work must be done on the origin of life before we can say it's inevitable or even probable, given a particular set of circumstances. "Possible" and "inevitable" are different; we must decide: What are the limits of possibility?

Ultraviolet Light

HULETT: Should we start talking about energy sources *per se* first? My personal opinion is there are several reasons to believe the ultraviolet radiation is probably the most useful energy source for prebiological synthesis. Obviously, there is a great quantity of it as you all know. It reacts preferentially with certain molecules. It can transfer energy quanta of about the right size to permit synthetic reactions, but the quanta are not so big that you have almost obligatory destruction, as there is with cosmic rays, x-rays, etc.

UV can penetrate below the surface of water, so reactions in liquid media can be speeded up by this energy source. This is an advantage over say, electrical discharges, which may be quite useful in some areas, but which I don't think can be used past the time when the initial products are made and put down into the water.

MILLER: But cyanide plays an extremely important role in prebiological chemistry and cyanide is most efficiently made by electric discharge. HCN is very hard to make by UV, with the possible exception of these hot hydrogen atoms from the photolysis of H_2S. There is no published work I know of where cyanide is made with UV.

SAGAN: It is in press now (Sagan and Khare, 1971).

MILLER: This is an exception. To the extent that cyanide is important electrical discharges are important.

HULETT: I agree.

MILLER: Secondly, the mechanism of generating atmospheric electricity is not really clear. A number of proposed processes have been suggested, in one the presence of traces of ammonia make a big difference. Charges are put on the surface of ice, as half a rain droplet, is frozen. I'm not sure whether more or less is generated, but this is peculiar to ammonia. It is a very complicated problem.

ORGEL: May we hear the mysterious discussion of making cyanide with UV?

SAGAN: Certainly. I referred to it this morning. There are several aspects. One approach is the use of H_2S as a long wave length photon acceptor, or in comparable experiments the use of formaldehyde as a long wave photon acceptor. In both cases amino acids are made in high yields. However, quantum yields are about 10^{-4} amino acid molecules per long wave length photon absorbed.

If cyanide must be an intermediary in the production of amino acids, then obviously we made it. In addition we have highly reliable (by the standards of this morning) matching high resolution mass spectra of acetonitrile in these experiments. Acetonitrile was alleged to be produced in there in experiments Stanley and I did with sparks eleven years ago (Sagan and Miller, 1960). For that reason I checked it out. Clearly we have made cyanides in these experiments (Sagan and Khare, 1971).

PONNAMPERUMA: What are your wavelengths?

SAGAN: 2537 and 1849 Angstroms only.

MILLER: Was there mercury in the system?

SAGAN: No, definitely not. We took great care to make sure there was no mercury.

HULETT: What concentrations of H_2S and formaldehyde were used?

SAGAN: Our H_2S to ammonia ratio was about one. This is consistent with H. D. Holland's (1962) geochemical expectation for the early earth.

We studied a range of compositions, but in each case we wanted H_2S to be optically thick in a container of laboratory size, therefore the absolute abundance of H_2S was probably large for the primitive atmosphere.

LEMMON: Was H_2S and formaldehyde always present?

SAGAN: No. In one set of experiments there was H_2S and no formaldehyde, and in another there was formaldehyde and no H_2S.

ORGEL: What other components were in the system?

SAGAN: Nothing, water vapor, ammonia, and ethane.

ORGEL: Were any experiments done with nitrogen instead of ammonia?

SAGAN: No. In experiments with no ethane, however, we did not find any detectable amino acids.

FERRIS: May I ask you a question before you continue? As Chairman, you may either answer it or not!

SAGAN: I'll try to answer it.

FERRIS: The quantum yield for the photolysis of hydrogen sulfide to H· and HS· is practically one. The eventual photoproducts of hydrogen sulfide are hydrogen and sulfur. My concern here is: Hydrogen sulfide is very rapidly photolyzed. For your process to be reasonable, the sulfur must be recycled somehow. Have you a mechanism for recycling the sulfur for reuse? To make this a reasonable prebiological synthesis, you must get sulfur back up to hydrogen sulfide —

SAGAN: Stanley raised this question with me yesterday, for the first time. I hadn't thought of it earlier; I'm not sure I understand the problem.

Are you worried about rapid conversion of H_2S to sulfur by ultraviolet light on the primitive earth and therefore depletion of the H_2S? Obviously there is some steady state concentration from outgassing and deposition. I don't know what it is, but surely there is some way of maintaining it, I just don't know the chemistry of reformation of H_2S.

MILLER: A possible mechanism is the reaction of sulfur with cyanide to make a thiocyanate, and then the hydrolysis of the thiocyanate to CO_2, ammonia and hydrogen sulfide. The hydrolysis would be a rather slow reaction. Someone should look at the chemistry of sulfur to see how to get back to H_2S. This is just one possible mechanism.

SAGAN: The point of these experiments is not that H_2S absolutely must be the long wavelength photon acceptor; rather there are several compounds (at least two: formaldehyde and H_2S) which absorb at long wavelengths and happen to give off a hot hydrogen atom when they photodissociate. The absorption of long wavelength photons leads to amino acid synthesis with a high quantum yield.

The long wavelength ultraviolet light, because of the shape of the Planck distribution, is by far the most abundant. The short wavelength flux starts very feebly and increases exponentially, gaining an order of magnitude with every increase of a few hundred Angstroms toward the Wien peak. If the long wavelength photons can be utilized a lot of energy can be picked up. That is the main thrust of the experiments: energy previously thought to be unutilizable may have been effective.

FERRIS: My concern was the same argument I have had with formaldehyde, which I think Dr. Hulett has also raised: How will enough formaldehyde be built up to do this?

Another question of concern is: What are your yields? Not quantum yields of amino acids, but conversion of, say, ammonia or nitrogen to amino acids? Give the ratio of nitrogen to sulfur can we consider how much of that nitrogen would be converted into amino acids?

SAGAN: It is something like 10^{-3} of all the sulfur atoms put into the system converted to amino acids.

MILLER: Do you mean cysteine?

SAGAN: Since the sulfur to oxygen ratio is unity, –

MILLER: The yields are usually done on the basis of carbon.

LEMMON: Did you specifically identify cyanide?

SAGAN: Yes, –CN was identified in the gas phase, but not HCN. There is a peak at that mass number in low resolution mass spectroscopy. The high resolution mass spectroscopy on that peak was not done, because we had produced CH_3CN.

MILLER: If cyanide is limiting, it would be nice to know your yields; and a silver titration is an easy way to determine the HCN.

SAGAN: I agree. I have your comment. We'll go back and do it.

PONNAMPERUMA: Haven't some of Philip Abelson's (Carnegie Institution of Washington) early experiments shown the presence of cyanide?

PONNAMPERUMA: I don't know all the details, but they identified HCN and formaldehyde in photochemical experiments.

MILLER: Was this from formate in UV in solution?

PONNAMPERUMA: I though so. I've forgotten the exact details.

MILLER: In terms of amino acid synthesis the old experiments of Groth and von Weyssenhoff (1957) have suffered from the fact that they were very low on the cyanide, although they probably made enough aldehydes. It would be interesting to do this with nitrogen.

SAGAN: Might you elaborate on your question?

ORGEL: The general notion is that it is difficult to accumulate large amounts of ammonia in the atmosphere because of photodecomposition but there is no corresponding difficulty with nitrogen. It seems to me that nitrogen should be in substantial excess over ammonia in any reasonable atmosphere.

SAGAN: This may be really a topic for a later discussion, but is your expectation consistent with the known composition of the atmosphere of the planet Jupiter?

FERRIS: We have some evidence. We studied the photolysis of ammonia using short wavelength light of 206 nanometers to study this particular question. Ammonia would very rapidly be converted to nitrogen except in the presence of a fair excess of hydrogen. We have used roughly four times the quantity of hydrogen to ammonia – decreasing the quantum yield for ammonia photodestruction from 0.25 to 0.01, which is about what we can measure.

MILLER: Hydrogen protects with respect to decomposition?

SAGAN: How much excess of hydrogen do you need

FERRIS: The experiment used 10 torr of ammonia, and we did it with various pressures (more recent data at 1 torr is also given in Fig. 7). This graph shows a very sharp break at roughly 40 torr of hydrogen.

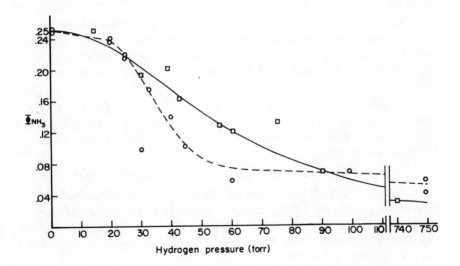

Figure 7. Change in quantum yield for ammonia photodecomposition with hydrogen pressure. The final pressure in the photolysis cell was adjusted to 740 mm with argon. (□ – □ – □ – □ – □) ammonia pressure of 10 torr; (O – O – O) ammonia pressure of 1 torr. Results of Dr. David Nicodem, Rensselaer Polytechnic Institute.

SAGAN: Does this imply that on the earth hydrogen is not needed as a major constituent of the atmosphere in order to preserve ammonia?

MILLER: There are two parts to this. First is the absolute pressure or the ratio important?

SAGAN: Do you mean the absolute pressure, or the absolute pressure of hydrogen?

MILLER: He says the break comes at 40 torr or millimeters. The ratio is 4 hydrogen to 1 ammonia. Is the 40 mm or the 4 to 1 ratio important?

SAGAN: The important thing is 40 mm of anything. Surely there were other gases on the primitive earth. If anything, it's 40 mm of hydrogen.

ORGEL: Is that consistent with the outer planetary observations?

SAGAN: On all outer planets there is much more than a 4 to 1 ratio of hydrogen to ammonia, which would be consistent.

If on the other hand the ratio is important, I can imagine that the primitive earth had a low ammonia abundance (because of its solubility in water) and the atmosphere had more hydrogen than ammonia, both of which are trace constituents.

MILLER: With a rate-of-escape calculation you come out with about 10^{-3} atmospheres of hydrogen, about a thousand times as much as the numbers Jeff Bada and I were playing with. If it's 1000 to 10, and the 4 to 1 ratio of H_2 to NH_3 is important, then ammonia would have been protected very well.

There's another part to this: Can sulfide be protected from decomposition by an excess of hydrogen? The literature data indicates that it can not, but no one has gone on to large hydrogen/hydrogen sulfide ratios. This might be worth doing.

ORGEL: Jim, what do you think happens with the $NH_2 + H_2 - $?

FERRIS: I don't think it's really the ratio, but rather the absolute quantity of hydrogen. I think there is a reverse reaction, and the primary step is NH_3 going to $N_2H\cdot$ and $H\cdot$, and we're just effecting a back reaction. By photolyzing ammonia in the presence of deuterium instead of hydrogen we see the formation of NH_2D.

MILLER: Are the bond energies right to split H_2: $NH_2 + H_2 \rightarrow NH_3 + H$?

FERRIS: They are very close, just about equal.

PONNAMPERUMA: It's back and forth. But what happens with methane in the system, if methane is photolyzed at the same time?

FERRIS: Here we're irradiating ammonia. The methane has no effect on the rate of photolysis of the ammonia.

PONNAMPERUMA: That's rather extreme, because methane will break down too, and you will get NH_2.

FERRIS: No! We are using essentially a monochromatic light source which will activate only the ammonia, not the methane.

PONNAMPERUMA: But what happens with a spectrum like the primitive earth atmosphere?

FERRIS: We haven't done this.

PONNAMPERUMA: This relates to the question that Carl asked. In the presence of methane — methane may be photolyzed, and provide enough hydrogen to maintain that.

SAGAN: Photodissociation of H_2S produces hydrogen by itself, and if there were reformation, the process would be itself self-limiting.

ORGEL: How about the photolysis of S_2?

MILLER: The photochemical literature indicates S_2 always polymerizes.

SAGAN: Our experiments show no significant mass spectrum peak at the mass number for S_2 in the gas phase, but in the solid phase we have all the sulfur polymers up to S_8, with the correct isotopic abundances.

MILLER: Because if S_2 can find another S_2, it forms S_4, and this process continues up to S_8 and makes a ring.

SAGAN: In the gas phase?

MILLER: Yes. The stable form of sulfur, monoclinic and rhombic is S_8; S_8 molecules with puckered rings are in the crystal. It is very difficult to break that ring. With the mass spectrometer S_8 is broken down by ionization.

USHER: As sulfur is melted, though, the rings were broken.

MILLER: Yes, but bottles of "flowers of sulfur" contain S_8.

NAGYVARY: What is the highest wavelength of light that can photolyze H_2S down to sulfur and hydrogen?

SAGAN: I don't know but since H_2S photodissociation at 2537 Angstroms leaves the hydrogen atoms with two electron volts superthermal, then it must be possible to photolyze it at wavelengths at least several hundred Angstroms longer than 2537.

FERRIS: Doesn't the spectrum go all the way out to 3000?

SAGAN: It's falls more or less exponentially.

The H_2S spectrum, asymptotically approaches zero, so it is very small out at 3000 Å. It would not surprise me if 2800 Angstrom radiation was still adequate for photodissociation.

We might wish to discuss other energy sources in comparison.

MILLER: People have assumed that the UV would end up in the ocean and do all sorts of wondrous and wonderful things, but it always has been my impression that the primitive ocean was very opaque, from for example: the yellow and black polymerized cyanide that comes from electric discharge experiments. This yellow and black tar has a very large absorption in the UV. Organic reactions with reactive materials – aldehydes particularly – generally give this yellow tar which is hard to get rid of. Although a nuisance for the organic chemist, it might be very effective in protecting against UV absorption by compounds in the ocean. It may have been very difficult to do UV synthesis in the ocean. For example, the tetramer of cyanide to the amino-azonitrile might have been impossible in the ocean, and may have been more probable in rain drops.

PONNAMPERUMA: I have a preprint of H. D. Holland's recent paper submitted to *Science* in which he has made some calculations based on the UV flux in the primitive atmosphere that suggest there may have been an oil slick one meter thick on the surface of the primitive ocean! The Santa Barbara oil spill pales into insignificance relative to this!

HULETT: If the oil slick is one meter thick, I'd have to agree with you! The first centimeter, or something may have had a very high concentration of yellow tar, yellow foam, or whatever. This brings up another point, and that is: In the shock synthesis experiments Sagan (Bar-Nun *et al.*, 1970) has done recently, I don't think they could have gotten much besides glycine and alanine, because the efficiency was extraordinarily high. If glycine and alanine were produced that way I doubt if other polymers would have been produced.

MILLER: Yes, but the absorption at 2500 Angstroms of the solution right from the electric discharge apparatus has an optical density of several hundred.

SAGAN: How much?

MILLER: I forgot the exact numbers, but it's very high. It has to be diluted to get a reading on the spectrophotometer.

HULETT: This brings up the question of density, compared to what it was in the atmosphere. How much of this polymer has been concentrated in that particular solution, compared to what would have been in the ocean?

MILLER: That's a somewhat different question. I can't estimate how much was around, but if there was much, it would have made the ocean very opaque in the UV.

SAGAN: Certainly, but after all, there is slow oceanic circulation, and the fluxes we are talking about are immense.

BADA: What do you mean by "oceanic circulation"? Mixing ratios of the order of a thousand or two thousand years? Do you consider that slow?

SAGAN: It isn't fast, but it does mix!

In the presumed primitive atmosphere in the wavelength region we are referring to (2000-3000 Å), CH_4, NH_3, H_2O, H_2, CH_2CH_3, CO, and N_2 are all transparent; it is the more complex molecules that begin to absorb. The ones with the strongest absorption coefficients that also have some probability of having been formed and survived are the aldehydes and H_2S. Ketones also absorb.

If I put in the absorption spectra of these materials, I find a window in the primitive atmosphere which tends to be around 3500 to 2600 Å. This is an interesting wavelength region considering the absorption of purines and pyrimidines. The point I want to stress is that the fluxes in this wavelength region are immense: Maybe 10^4 ergs per square centimeter per second. Obviously a great deal of photochemistry can be done with those fluxes, even if attenuated by a factor of 10^{-4}.

I imagine there are molecules absorbing the huge UV flux at the surface of the ocean, which are then circulated down and you don't see this radiation source, as Bob says, for thousands of years. The material synthesized by the UV flux is then protected from subsequent photodissociation.

MILLER: Exactly.

HULETT: Are the yellow tars photolyzed when they absorb? What happens to them?

MILLER: That was never studied.

PONNAMPERUMA: That is a very worthwhile experiment that nobody has done. Some of the UV might be absorbed by this muck floating on the ocean, and form some very interesting products.

MILLER: Have you ever measured the absorption coefficient of the yellow material from the cyanide polymerization?

SAGAN: We have measured the absorption of the yellow material in our stuff which I assume is part cyanide. It's so opaque already in the blue and the ultraviolet that it doesn't go up orders of magnitude into the ultraviolet. Just by looking at it with a blue filter you can get an idea of its very large opacity.

MILLER: Then would not the UV light in that window get absorbed in a micron or a centimeter of ocean?

SAGAN: Surely a centimeter of this material will reduce the UV by many orders of magnitude, and because of the exponential statistics, doubling the amount of material will cut down UV penetration by another many orders of magnitude.

MILLER: Right.

SAGAN: So I don't doubt that a meter thickness of this sludge prevents all respectable photons from getting through to the other side.

SIEVER: Why do you believe that the polymeric material keeps growing on the surface of the ocean? Recent work on oil spills tends to show that their disappearance is not all oxidation but that it actually starts sedimenting. It increases its density and falls to the bottom. This explains the fact that oil slicks of the world are not still all around the surface of the ocean.

What is the current cause of the increase in the density of these tarry materials? I don't know what the brownish stuff would do, but I think it is very possible that it would sediment. The process would be cyclic: it would be made, decompose, in the sense of changing its density, so that it would no longer be equivalent; or it might be a sort of steady state filter.

MILLER: This tar for the most part is water soluble. I wasn't speaking of an oil slick, but of a soluble polymer.

SIEVER: Then it would be mixed.

MILLER: Yes. You don't need Holland's oil slick to do it.

SIEVER: There would at least be a mixture of several hundred meters of the ocean; this would reduce its opacity immensely.

HULETT: It's generation rates versus destruction.

ORGEL: What difference does it make whether all the light is absorbed in a millimeter —

MILLER: If you are worried about UV destruction of the compounds that you have worked so hard to synthesize, the narrower the better.

HULETT: On the other hand, if you want UV synthesis of polymers, then the wider the better!

MILLER: Wasn't it your thought, Bob, that photolysis might occur in raindrops? This looked nice for the rearrangement of cyanide tetramer to amino imidazole nitrile.

SANCHEZ: Yes, I think that is most likely.

ORGEL: The surface of ice is another good place.

BADA: The average depth of the ocean now is 4000 meters. We don't know what it was 3.5-4 billion years ago, but it may have been 2000 or so. How much UV light penetrates, even to a depth of 25 to 30 meters? I think it is essentially removed; most of the organic material will be out of the influence of UV radiation in the ocean.

SAGAN: At about 2600 Å tens of meters thickness is one optical depth in pure sea water.

NAGYVARY: What happens to sulfides in water when UV irradiated? Is it the same as irradiating H_2S in the atmosphere?

SAGAN: Since the liquid water bath that we have in our experiments never sees the ultraviolet light, we have not done that experiment!

MILLER: Joe Nagyvary raised another point: sulfide in water absorbs like sulfide in the gas phase, but the HS ion in water absorbs at 2600 Å. The absorption coefficient seems pretty large and at pH 8 most of the sulfide would be dissolved in water as HS^- —

NAGYVARY: If it reacts with sulfur, probably it forms polysulfides.

SAGAN: Yes, we make polysulfides.

NAGYVARY: Polysulfides absorb at much higher wavelengths, at or above the 300 region. If there was enough sulfide and polysulfide in the ocean, then the protection against UV would have been much higher than by those yellow tars, because of the great difference in the UV absorption. It would be of some interest to examine if the sulfides in the water could participate in energy transfer to have certain reactions.

FERRIS: Other than sodium and potassium most sulfides that I am aware of are quite insoluble. This concerns me about the use of hydrogen sulfide (you may correct me) — but why wouldn't any sulfide in solution immediately precipitate out as some insoluble metal salt?

MILLER: Yes. Ferrous!

SIEVER: Nice and black.

MILLER: There are two possibilities: With an excess of sulfide there would have been very little dissolved ferrous ion; with an excess of ferrous there would have been very little sulfide. From the standpoint of the phenylalanine synthesis, we needed sulfide, so we postulated an excess of sulfide. I don't know if this is a valid point.

FERRIS: Would one of the geologists say something about this?

SIEVER: You are right. There probably was a lot of ferrous iron around — no, not much ferrous iron, because there would have been ferrous hydroxide in an ocean, with a pH of around 8. Since the pK is relatively low, ferrous hydroxide would be limiting.

ORGEL: Would there be an exchange?

SIEVER: Yes, there would be FeS at the lower pK and presumably a lot of black ferrous sulfide.

SAGAN: The cosmic abundance of iron is almost the same as the cosmic abundance of sulfur, so it is very hard to predict which would be in excess.

NAGYVARY: There are very great concentrations of sulfur in certain localized areas in Texas, for example, where there is no great abundance of iron.

SIEVER: That is all the product of bacterial activity.

ORÓ: Texas sulfur is not applicable in this case. We have raised an important preplanetary problem: if there is no meteorite chemistry the sulfur on the primitive earth was probably already in the form of sulfides of iron and other metals. On the moon, apparently, most sulfur is in the form of iron sulfide. The sulfur in most of the meteorites we have analyzed is also in iron sulfide.

The sulfur required for prebiology must come from the composition of these sulfides at the formation of the earth from the escape of internal hydrogen, or some water reaction at high temperatures, and then released as today from volcanoes. The sulfur found today in sediments is actually biological sulfur, produced by sulfate-reducing bacteria. It does not apply to the over-all problem, except that in the carbonaceous chondrites some free sulfur is present. Still there is no hydrogen sulfide, ammonium sulfides, and no sodium sulfide as far as I know.

SIEVER: We don't know their pK's but the ferrous silicates actually might be more controlling, and therefore allow the mobility of sulfide. The pK's must be very low.

MILLER: Isn't R. Garrels (pers. comm.) starting to measure these things?

SIEVER: Yes, he is, I am, and various others are, but no one has any good numbers yet.

SAGAN: May we continue with the discussion of energy sources? We have gotten into a sulfurous backwater. Shall we continue with UV or discuss other energy sources?

Thunderstorms

HULETT: Another item on ultraviolet efficiency is that it can release its energy essentially completely at a surface, whether the surface of the ocean or a solid-liquid interface or an air-solid interface. Surface catalyzed reactions which require energy inputs are much easier to get from ultraviolet than, again, from electrical discharges or some other sources where you have a large volume of material into which the energy is put.

Can anyone comment on that? I think most other energy sources are generally volume sources. A heat source, such as a volcano, is not going to be concentrated on a particular area, but will be over some volume of lava that is exposed. Energy generated by meteorite impact is something else. A volume is swept out by the meteorite as it enters; whereas with any photon source all the energy can be released at a spot where it can do the most good.

SAGAN: Will you discuss the nonultraviolet energy sources in some detail?

HULETT: I would just like to get your comments on thunderstorms (I'm sure you know more about it than I do); but one question concerns what the temperature was on the primitive earth.

SAGAN: I hope to talk about that tomorrow. We have a set of time-dependent greenhouse models of the earth which involve the evolution of the sun. Anyway, I think the temperature is moderate to low.

HULETT: What I want to say may be modified by what you say to-morrow, but I have the impression that the sun was less luminous at the beginning of biological evolution than now. The temperature therefore was colder, unless there was a more pronounced greenhouse effect. Maybe there was; maybe there wasn't. If the same greenhouse effect as now is assumed, the mean temperature probably was 20 or so degrees less.

If so, much less electrical energy than now would be expected because electrical energy is a function of evaporating water, and warm, moist, humid air at the bottom, with cold, dry air aloft. A great quantity of water vapor must be transported up to the upper atmosphere to produce appropriate conditions for a thunderstorm.

At present a temperature difference of about 30°C changes the average number of thunderstorms, very roughly by something between one and two orders of magnitude. If so, considerably less thunderstorm activity would be expected on the early earth than now.

MILLER: I'm a little confused, I don't know their height, but thunder clouds are rather close to the earth; water vapor does not have to be trans-ported a large distance.

The limiting factor involved in the generation of thunderstorm elec-tricity, whether water transport or the charge generation mechanism is not really clear.

HULETT: To most meteorologists, it is quite clear that the problem is to transport water up; this involves a lot of energy available for release as vaporization of water. This means warm, moist air. For example, there are not very many thunderstorms in the Arctic, because there is not enough energy to evaporate the water.

MILLER: I think that the state of knowledge of thunderstorm elec-tricity in the present atmosphere is pretty limited.

HULETT: It's dim, certainly.

MILLER: I don't believe anybody can really say anything about what would happen under prebiotic conditions.

HULETT: Although I don't know how relevant this is, I can say some-thing about the ammonia concentrations in present-day rain water. The con-centration is highest around 45° N. latitude, and then it tapers off and be-comes lower down toward 45° S. latitude, because there is more organic pro-duction in the northern hemisphere.

The amount of ammonia is lower by about a factor of two in total amount, and by a factor of about three or four in dissolved ammonia in rain water in the tropics, as compared to 45° N. latitude. There are a considerably higher number of thunderstorms in the tropics than at 45° N., in spite of a lower ammonia concentration.

If there were a much higher concentration of ammonia, it may have more of an effect, but with this variation it doesn't have much effect.

BARGHOORN: Are you subtracting out nitrate production as a source of ammonia?

HULETT: I'm just discussing ammonia found as ammonia gas or ammonium ion.

BARGHOORN: In rain water?

HULETT: In rain water.

BARGHOORN: But much of this could be from nitrate reduction by bacteria.

HULETT: Yes, that's probably where it comes from initially. This is ammonia as it comes down in rain, so therefore it was ammonium ion at the time the electrical processes were taking place.

PONNAMPERUMA: Is there general agreement on that?

HUBBARD: I don't think so.

SAGAN: Is it generally agreed that low temperatures imply less thunderstorm activity?

PONNAMPERUMA: No, that is understandable but with the early low temperature of the earth are all the oceans considered to have been frozen?

SAGAN: No. I will argue that in detail tomorrow.

PONNAMPERUMA: Dr. Hulett's argument is based on low temperature, isn't it.

HULETT: Yes.

SAGAN: There are convincing arguments that temperatures were a little lower on the primitive earth than today. I will argue that they were not as much lower as expected in terms of the present greenhouse effect, though they were lower.

MILLER: Central to the discussion is the necessity that the temperatures were low; otherwise organic compounds would have decomposed. Obviously the lower the temperature, the more stable the organic compounds and the lower the decomposition. Low temperatures are needed for this reason.

PONNAMPERUMA: But everything didn't freeze out, either.

MILLER: This is one of these general arguments; the validity of it is another matter.

SAGAN: Let us please postpone the discussion of the temperature of the primitive earth until tomorrow when we will have more inputs.

Heat and Geothermal Activity

BADA: Wouldn't outgassing contribute water vapor to the atmosphere?

HULETT: Outgassing is much, much less than evaporation of water from the ocean. Apparently about 35 per cent of the solar energy that reaches the earth now goes into evaporating a tremendous amount of water.

SAGAN: There may have been much more outgassing early. The atmosphere and oceans had to come from somewhere.

MILLER: What is the time scale for evaporating the whole ocean with solar energy?

HULETT: About 10,000 years. It's fairly short.

SAGAN: An important distinction among the various energy sources is the extent to which a synthesized molecule can readily escape the energy which just synthesized it. Some energy sources like thermal energy clearly don't fit that criterion. Even if thermal energy sources were extremely efficient, and even if geophysically plausible (which I don't believe) still once a molecule is synthesized, there it is sitting in the heat that made it.

MILLER: But you might have had a pyrolysis of atmospheric gases, swept by wind, past the molten lava.

PONNAMPERUMA: An ocean may be next to your hot lava.

SAGAN: Right. A rainstorm may wash those molecules out to the sea, and so on; but in general something special must protect molecules from the energy source that produced them.

HULETT: The wind will not become heated passing across the hot lava. It will warm up and rise, rather than remain at the surface and get up to about 1000°.

MILLER: I like pyrolyses that involve very short contact times.

SAGAN: Yes, you get around the problem by having the heat for only a very short time. But that's relevant for only a tiny fraction of geothermal heating events.

MILLER: The experimental arrangement is to have the gas molecules, methane, or whatever, in contact with the heat source for a tenth of a second, or one second. If left longer than that, the product may be wrecked.

HULETT: The question is whether in the particular situation there is contact long enough to heat up the gases high enough to pyrolyze.

MILLER: I see.

HULETT: The air mass will move away far enough if there is a natural situation so that it doesn't really get up close to the temperature of the 500° or 1000° lava.

MILLER: The gas is actually in contact. The gases right next to a bed of lava are going to get heated.

HULETT: Yes, certainly, gas molecules in contact with the lava will heat up.

MILLER: But at a distance of not more than a centimeter, and probably much less.

PONNAMPERUMA: It's not so far-fetched. Take Surtsey [Iceland] for example, a classic example where the lava flow is in contact with the ocean. If there were many Surtseys around in prebiotic times, there must have been synthesis due to thermal energy. Do not invoke the water from the heavens!

SIEVER: But what is seen at Surtsey in detail is hot lava, and then quenching.

HULETT: Lava cools pretty quickly.

PONNAMPERUMA: Take the case of the synthesis of nitriles. Methane and ammonia come out of the volcano, the gases pass over the hot lava –

SIEVER: The time scale for quenching is important: how many minutes, hours, or days are there before the lava moves in and quenches immediately to a temperature of 15°?

PONNAMPERUMA: It is not any different from the chondrite on impact.

SAGAN: Ways of getting the stuff off the hotplate can be thought of and then we will argue about the geophysical plausibility of such events having occurred at high frequency. There are other energy sources for synthesis which have the saving mechanism built in and where we don't have to invent something to save the molecule. For example, electrical discharge – lightning doesn't strike twice, you know. This is relevant. If the molecules are synthesized during discharge, they are there after the stroke has passed by, and are not in the high temperature field produced by lightning.

PONNAMPERUMA: Thunder might break them.

HULETT: Lightning does strike two or three times in the same place.

LEMMON: In a billion years, yes.

PONNAMPERUMA: Sonic boom.

Shock Waves

SAGAN: Shock waves are even better than electrical discharges, in this respect, because they last even less time than electrical discharges. As a pressure hypervelocity shock wave passes through a certain spot, there are instantaneous high pressures and temperatures. Then quenching occurs extremely rapidly. This seems to me a reasonable explanation of the high yields produced in shock synthesis experiments (Bar-Nun et al., 1970   ).

Ultraviolet light is less satisfactory in this respect. If, in the atmosphere the molecule synthesized is larger than the molecules you started from (if not, we're not concerned with it) it obviously is more likely to have a large absorption coefficient. As molecules are synthesized it becomes more and more likely that they will be destroyed.

That's why it's nice to imagine UV synthesis near the surface of the oceans. I like long wavelength UV, because only it will reach down near the surface of the water.

WOLMAN: Something really has bothered me since I read your paper on the short wavelength synthesis. Maybe we can put data on this. You wash out the tube with dulute hydrochloric acid, and put down the stuff on the analyzer, and got your amino acids. Am I right?

LEMMON: Be careful!

SAGAN: Two methods were used, both paper chromatography and the amino acid analyzer.

LEMMON: How about that HCl?

WOLMAN: I'm not discussing the HCl now.

I would expect, if you get something, it would be similar to the electric discharge results, or to those using HCN. I would expect aminonitriles, and not amino acids. The aminonitriles, before being put together and on the analyzer, or on paper chromatograms, they would have to be hydrolyzed.

SAGAN: Yes. It is known that when the material hydrolyzed, the amino acid yield is increased.

WOLMAN: I'm just asking. If no acid hydrolysis has been done, I'm a little bit surprised that you found free amino acids.

SAGAN: In the UV experiments we get amino acids before we acid hydrolyze. We get much more after acid hydrolysis.

WOLMAN: Did you compare what you got before and after acid hydrolysis.

SAGAN: In the UV work, the yield after acid hydrolysis is about an order of magnitude higher than before acid hydrolysis. In the shock experiments no acid hydrolysis was performed.

MILLER: It has never been repeated?

SAGAN: Akiva Bar-Nun is in Jerusalem now, drumming up to do a whole set of experiments.

USHER: How long were the products in contact with the HCl before analysis of the amino acids?

PONNAMPERUMA: There could be some mild hydrolysis in the water.

MILLER: Do you remember how the Bar-Nuns got rid of the HC1? As I remember the experimental arrangement, the products were swept into a glass round-bottom flask containing 20 or 30 ml of HCl. Surely they didn't put the aqueous HCl on the analyzer.

SAGAN: I understand you but I don't remember the answer. The experiments were done a year and a half ago.

ORGEL: I remember reading the paper, and being struck by the absence of nitriles, and presence of acids, had conventional wisdom been right. It's puzzling and suggests, perhaps, a completely different mechanism.

SAGAN: I quite agree that it's important.

BADA: Did you see isoleucine?

SAGAN: Dr. Hulett, have you a copy of that paper with you?

HULETT: Yes. Do you want to take it?

MILLER: I remember there was valine, but not norvaline.

BADA: If you make isoleucine, you have to make alloisoleucine. That's the key to whether it is real, or contamination.

SAGAN: No, there's no isoleucine.

BADA: Is there threonine?

SAGAN: Only four amino acids appeared both in the column and on paper: glycine, alanine, valine, and a trace of leucine. The relative proportions are: glycine, 73 arbitrary units; alanine, 35; valine, 0.5; and leucine, 0.1. So it's almost entirely glycine and alanine.

MILLER: Carl, isn't it very strange to get valine but not norvaline, which is the straight chain?

SAGAN: I have no idea what the kinetics are in shock synthesis.

ORGEL: Had there been nonnatural amino acids, would they have been commented upon?

SAGAN: Sure! The same instrument, the amino acid analyzer, had been used for all sorts of nonproteinaceous amino acids.

ORGEL: Because, again, there are a lot of isomers that one would have anticipated.

SAGAN: I agree, it is important to redo this. I'm sure Bar-Nun knows the answer to your question. Unfortunately, I don't. Maybe we can ask him in time for the publication of this conference transcript. That surely will give us several years.

HULETT: As far as amino acids and HCl are concerned, the yields they got were high enough so that it was much —

WOLMAN: No! I'm not saying these amino acids come from HCl contamination. I am just wondering about the whole mechanism of this reaction.

HULETT: The shock synthesis was just amazingly efficient. Looking at the amount of energy that could have gone into the molecules that made the amino acids, essentially 100 per cent of it went into making those amino acids.

SAGAN: Some tens of per cent. There's very little energy left over. We get 5×10^{10} amino acids per erg absorbed, a yield of four orders of magnitude in the same units greater than in the UV experiments.

MILLER: Carl, did you ever think that the efficiency is so great something might not be strictly kosher?

SAGAN: Yes. We did all the obvious controls: the exact procedures in the same shock tube, but without the shock.

PONNAMPERUMA: We have had difficulties with shock waves. Together with Adolf Hochstim, now at Wayne State, who did some of the initial calculations on meteorite chemosynthesis, we have recently tried other sorts of shock waves. His graduate student, Woo Park, tried to reproduce the shock generated by a laser beam (10 kilojoules). We know the shock is there, because the fireball inside the apparatus can be photographed, but we couldn't detect any amino acids.

MILLER: When you say "we couldn't detect any" what upper limit to the per cent efficiency is implied?

PONNAMPERUMA: By gas chromatography, for example, 10 nanograms of amino acid synthesized would have been detected.

SAGAN: Your sensitivity is obviously very high, but what kind of a shock was it? Have you a continuous laser beam?

PONNAMPERUMA: No, a pulse laser.

SAGAN: What is the pulse frequency? It's milliseconds for the shock wave.

PONNAMPERUMA: A microsecond in this case.

SAGAN: The pulse is on for a microsecond, and then it is off, is that correct?

HULETT: How well does the kind of energy and the type of shock wave created simulate a natural energy source?

SAGAN: Very well. The shock waves produced in thunder are particularly well simulated, because of the millisecond or so at high temperatures: 1000 to 2000 Kelvin.

ORGEL: How do the efficiencies in these experiments compare with those in totally unrelated, nonprebiotic reactions?

SAGAN: I think very well.

PONNAMPERUMA: They should be, this is one of the bases for much of the work. Hochstim's calculations predicted very high efficiency (Hochstim, 1963).

MILLER: Perhaps shock waves are efficient for making acetylene, but are they efficient for amino acid synthesis?

PONNAMPERUMA: He didn't make calculations for organic molecules, but rather for the dissociation of the gases. He assumed that recombination would give rise to many kinds of organic compounds. Hochstim's figures, predicted a meteorite 11 km in diameter, traveling at a rate of 11 km per second, would yield something like 10^{12} tons of organic compounds!

SAGAN: I did a computation on this once. Shock synthesis gives a very high efficiency. There's every *a priori* reason to think that shock organic synthesis is highly efficient.

MILLER: Cyril, these are theoretical calculations, and I think Leslie is asking: if a shock wave passes through methane, for example, how many moles of acetylene per kilocalorie is obtained?

PONNAMPERUMA: Carl, Bauer, your associate, should have that.

MILLER: It might be very high, but I doubt if you would get 30 per cent. This is just a guess, but one ought to know for sure.

ORGEL: There is also the suggestion that this method may have something to do with HCN.

SAGAN: Who suggested that?

ORGEL: I think the idea went around the table. If all of the stuff in the shock wave was turned into HCN, there would be wastage through polymers, I suspect.

PONNAMPERUMA: HCN was used as a criterion of comparison.

ORGEL: This synthesis would have to be by a mechanism quite different from the HCN mechanism; it seems too efficient.

SAGAN: My discussion was in the UV context, not the shock context.

PONNAMPERUMA: It may be wrong to assume that shock wave amino acid synthesis is via the nitriles; this may explain why Carl is getting free amino acids.

SAGAN: Free amino acids were made, and I quite agree that it's inconsistent with Strecker synthesis. Alternative synthetic pathways may exist.

There probably is no contamination in the system. We did a large number of controls. The relative abundances of the amino acids are not consistent with biological contamination.

MILLER: Afternoon coffee is ready. Perhaps we should continue this discussion later.

— — — — — —

SAGAN: Dr. Hulett, would you please continue with your remarks?

HULETT: I think we probably ought to discuss meteorite organic syntheses by shock waves a little more, because of the fact that Carl is involved in it. I object to the idea of much net synthesis by meteoritic shock, because at least in the current atmosphere their energy will be released in a part of the atmosphere where the density is very, very low and where the temperature is very high. The net result is that anything synthesized is probably destroyed in a fairly short period of time; certainly before it arrives at the surface.

SAGAN: Micrometeorites will certainly be thermalized at high altitudes, about 110 to 90 km above the earth's surface. Synthesis at such altitudes is not likely to be very efficient, because it takes a long time by any process for material produced at these altitudes to reach the surface. Neither will there be any atmospheric shielding from ultraviolet light at those altitudes. I think this is a reasonable conclusion for the primitive earth.

Larger meteorites come down deeper, and some celestial objects of perhaps millimeter size and even larger make it to the surface of the earth. There's an exponential decay in their distribution function. Less than a per cent or a tenth of a per cent of the meteorites which enter the atmosphere come down deep.

But as we said in our paper (Bar-Nun *et al.*, 1970) the vast majority of the shock energy on the earth today is from thunder and not from micrometeorites. I would expect that on the primitive earth there were a lot more meteorites, and somewhat less thunder.

HULETT: I would also like to ask what happens after there are monomers, in the sense that most of the rest of the synthetic reactions require an input of energy. They are not going to happen just because there is a catalyst there. Reactions won't happen just because there is a relatively high concentration. Probably a specific input of energy into the synthetic processes will be required. How do you envision prebiotic synthesis of polynucleotides? Was it necessary to have polynucleotides? Was there a primitive system that didn't depend upon the template formation we have now?

MILLER: What do you mean by specific energy input?

HULETT: I mean energy which permits the formation of a particular bond. A good example is the shock synthesis that Sagan has discussed. It seemed as if most of this energy went into the bonds forming glycine and alanine. If there is another system to form polynucleotide bonds, the situation is much simpler than if there is just a general energy input where a small percentage goes into —

MILLER: Does cyanogen fill the bill? It is of greater or lesser efficiency at synthesizing pyrophosphate bonds and polymerizing nucleosides.

HULETT: What experimental results are there trying to get polynucleotides with cyanogen?

Prebiotic Synthesis of Polymers

MILLER: Leslie and Bob Sanchez have done these experiments.

ORGEL: None are very good. For prebiotic molecules of cyanogen halides, I think you get up to 10 or 15 per cent — is it? Let us explain the experiments.

LOHRMANN: With cyanogen we can phosphorylate the solution of nucleotides, but we are not able to get the nucleotide linkages with this. This was tried on 5' — uridine phosphate plus uridine. It didn't give any UPU. With inorganic phosphate you get phosphorylating within the system.

It was also tried on a template, where the efficiency should be better. There was no UPU formation. Neither were any inter-nucleotide linkages formed.

USHER: That was with 5' phosphates?

LOHRMANN: Yes.

MILLER: But cyanogen chloride was very interesting as I remember.

LOHRMANN: The reaction was done in the presence of imidazole, the cyanogen chloride reacting on an intermediate, very efficient for phosphorylations.

MILLER: Is this what you mean by a specific energy input?

HULETT: I mean something that will direct the energy into a particular —

ORGEL: There are, by and large, two widely accepted models it's very hard to choose between. (Many of the people here have played some part in the development of the present position). Stanley and Rolf have described one in which energy goes into a particular high energy chemical compound, with dehydration. The other model is much more sophisticated; things are just heated together, both react somewhat. I would put my money on heating things together at the moment, but it is obviously a very hard choice, and neither works extremely well.

We have done a lot on this. Cyril, you have too.

PONNAMPERUMA: The idea of building up high energy molecules, and proceeding from there is much more acceptable chemically.

ORGEL: Thermally?

PONNAMPERUMA: Building up high energy molecules by any method.

ORGEL: On the primitive earth were monomers made and then heated or is it more likely that monomers were made and something fancy happened to them?

SAGAN: Not a fair question. [Laughter]

MILLER: We haven't enough information to decide on this. Obviously it could be both.

PONNAMPERUMA: From information we have, obviously, the more general solution may be more acceptable. Thermal energy must have been available only in isolated microenvironments. I prefer to see reactions in the ocean rather than on the sides of volcanoes.

ORGEL: I like to see things happening in La Jolla!

NAGYVARY: There are fundamentally three ways of making poly-nucleotides; chemically or thermally activating the phosphate; phosphate must be activated to form pyrophosphate, which is very unstable in water. Therefore I don't like that method particularly, but it could have played a role.

Secondly, the carbons may be activated. Activated carbons are more stable in water; but how is the carbon activated? This is not easy to decide, but displacement reaction is another possibility. Thirdly, (this both Leslie and I think is most likely) is to raise the energy level by the formation of a mono-meric diester (which might or might not be a cyclic compound) and then make transesterification reactions. The most promising lab experiments to date are of this sort of chemistry.

If we involve thiophosphates in prebiological thinking, we might have a reasonably good mechanism to make diesters. If we have mechanisms to form diesters which are reasonably stable, we might get closer to understanding how polynucleotides might have formed.

PONNAMPERUMA: We should consider polynucleotides separately from considering the formation of polypeptides. If we can synthesize activated amino acids, we perhaps might make polypeptides readily.

HULETT: Polypeptides have been synthesized (Fox, 1965, 1970).

PONNAMPERUMA: I don't know whether it's right to call Fox's polymers polypeptides.

HULETT: It's a step in that direction.

PONNAMPERUMA: Polymerization takes place but the arrangement of molecules in some larger units is not known, enought to say they are poly-peptides. I'm not even sure whether there are always peptide linkages.

FERRIS: As Dr. Hulett has implied: the real problems is how will any polymers be built up? As the molecules get bigger, with more and more bonds a steady state is reached where the rate of decomposition approaches the rate of synthesis.

I don't see a way out of this dilemma: how will anything of any appreciable weight be formed?

NAGYVARY: We need an as yet unknown mechanism permitting an extremely high rate of synthesis. No syntheses involving pyrophosphates or triphosphates without enzyme catalysts are fast enough to avoid degradation. Most phosphate activation-type reactions lead relatively rapidly into an equilibrium.

FERRIS: The equilibrium indicates that the monomer is certainly the stable form, versus certainly, many polypeptides.

ORGEL: The problem would have been solved if there were a method to synthesize for example, an aminoacyladenylate. I'm sure you know the experiment that Katchalsky has done: the aminoacyladenylate is added to a suspension of montmorillonite and polypeptides of over 50 amino acid residues long are made   (Paecht-Horowitz *et al.,* 1970).

PONNAMPERUMA: But he has had difficulties. That can be done with alanine, but not with other amino acids.

ORGEL: He can also do it with methionine, I know.

PONNAMPERUMA: The last time Chaspey spoke to us, he could make polymers easily with alanine, but not now with the others.

ORGEL: I think the bifunctional amino acid may cause problems; but I'm referring not to the specific method, but to the fact that in this experiment on clay there is a known way of getting over this problem. The great difficulty is that the intermediate he uses is not one that anyone knows how to make prebiotically.

WOLMAN: What about the stability of the polymers? Would they survive a long enough time to accumulate − ?

ORGEL: He claims they are absorbed extremely strongly on clay. This seems to me possible. This is at least one example of a method of obtaining large molecular weights and getting around the problem that polymers fall apart faster than they can be made.

SIEVER: You are suggesting that a template of some kind is necessary for polymerization?

ORGEL: I don't think it's a template, just that a high concentration is maintained.

PONNAMPERUMA: The clay may remove them from further destruction: if molecules are adsorbed on the clay surface, this prevents them from being destroyed.

MILLER: When adsorbed on the surface, the products can still decompose in various ways.

MARGULIS: Does he use heat or radiation, or is this something spontaneous?

ORGEL: You merely take the plate and shake it up. There's very little hydrolysis.

PONNAMPERUMA: Phil Banda in our lab has done several related experiments. Small units, suggesting a certain order can be made, but the large polymers can not. The large polymers can be made only with alanine, but maybe the smaller ones can be made with other amino acids too (Banda and Ponnamperuma, 1971).

Maybe some method of coacervation in the water is involved.

USHER: We should distinguish two probable steps: the formation of polynucleotide not directed by a template, not actually reproducing an information content (that may be just random thermal polymerization) and the reproduction of polynucleotide sequences. Clearly high temperatures are not useful for that. For the formation of complementary nucleotide sequences, one needs rather more sophisticated solution experiments, such as those they (Renz, Lohrmann and Orgel), are now publishing.

I'm not worried by the stability of RNA, because a double helix (of complementary strands with base pairing complementarity which you are familiar with) could have a very long lifetime, even though its synthesis could be quite slow (Usher, 1972).

RNA possibly was the initial polynucleotide, because it can be degraded and sequences can be scanned using preformed polynucleotides which maintain their phosphates.

The ribonucleotides used shouldn't be the 5' phosphates. Preformed RNA degraded down to monomeric units will degrade to a 2', 3' mixture if it's ordinarily attacked by the neighboring hydroxyl group. If these units are then used again, why not have started with them in the first place?

ORGEL: You imply the only reasonable precursor for RNA synthesis is the nucleoside 2', 3'-cyclic phosphate! [Laughter]

SAGAN: Usher is shaking his head!

USHER: The recyclization must be achieved so a constant source of energy is needed.

ORGEL: It is worth making this point specifically: there are really only two compounds to start with, because there are only two places the phosphate can be: the 3', or 5', A 5' can only be used once, because it will dephosphorylate to the 2' and 3'. Therefore it would have to be dephosphorylated and rephosphorylated before use could be made of it again as a 5′. As long as it stays 2′,3′, it can be hydrolyzed, closed and rused. It is more complicated than this, but presumably nucleotides got big. Life started with stuff that has to use more stuff that could probably most easily have gone into 5′, rather than 2′,3′. My own guess would be that both were important.

Now I'd like to consider the problem of equilibrium. Unfortunately, the 2′,3′ isn't a very highly activated phosphate, comparing equilibrium constants. I have worked rather carefully through data from the biological literature. The equilibrium constant is of the order of unity.

USHER: At what pH?

ORGEL: The range studied includes what we like to think of as reasonably prebiotic: between 7 and 9.

USHER: Equilibrium between what? Because he got a bimolecular reaction in one direction —

ORGEL: The reaction of a cyclic phosphate plus a polynucleotide to give a polynucleotide one longer has an equilibrium constant of one.

It's very hard to make long chains in solution. I don't see the way out of this problem. Doing the calculation using these figures, less than 1 per cent of this could be of length 10.

Perhaps 1 per cent of length 10 is good enough: but in a really concentrated solution of a nucleoside cyclic phosphate (assuming a clever prevention of hydrolysis and waste by opening) at equilibrium less than 1 per cent would be present in chains 10 monomers or longer.

USHER: Did you take into account the problem of entropy?

ORGEL: No. That helps. Using the data on the entropy of association you can calculate that you will get some, but not an enormous advantage.

USHER: Why only the entropy?

ORGEL: Entropy calculations, together with melting points are used to calculate what the free energy would be at different temperatures.

When you do this, it turns out not to help much.

USHER: Like how many?

ORGEL: We haven't done this carefully, but I doubt whether there would be 1 per cent in chains of 20 monomers. It depends very much on the temperatures, the lower the better.

PONNAMPERUMA: Carl, I know we are discussing the question of polynucleotides now, but according to Dr. Hulett, we wouldn't even have hydrogen cyanide to make amino acids.

MILLER: Back to Dr. Hulett's paper: on page 68 you say the low efficiencies of nonspecific energy flow implies even processes requiring only two or three steps will be expected to produce negligible amounts of final compounds. I don't feel that's a fair statement. We don't know a polymerization process that is satisfactory in all respects, but even so from our achievements so far we can get more than negligible yields.

HULETT: Where do I say that?

MILLER: At the bottom of page 68 and onto 69. I was struck by the word "negligible."

HULETT: "Negligible" is a matter of definition, but earlier I arrived at numbers like 10^{-6} or 10^{-5} molar, which would then have to react to give a third compound. By the time the second or third compound is made in the absence of specifically directing energy toward the synthesis of these particular compounds, very much wouldn't be left.

MILLER: But we have examples of productions that are more or less specifically directed. If this objection is not completely overcome, it has less force than it did when you wrote this paper.

HULETT: Okay. I was talking about a nonspecific situation. If there is a specific step, then obviously the equilibrium can be changed. There have been some cases since this was published where specific steps apparently are involved — and we are better off then we were two years ago.

ORÓ: I think papers like this are extremely valid. It has helped people that may not be familiar to focus attention on the particular problems. Now Dr. Hulett should write another paper suggesting possible solutions for the problems he has raised! [Laughter]

HULETT: If I had the possible solutions, I'd be very happy to write a new paper!

ORÓ: The last general problem: how do we build polymers? has been in the minds of all the people working in this field for as long as we have been working in it.

We are reasonably in good shape in terms of building monomers. Most of the important monomers, if not all, have been synthesized, not only amino acids, but also purines and pyrimidines. Although obtained in very small yields, the synthesis of thymine has also been accomplished recently in my laboratory (Stephen-Sherwood *et al.*, 1971).

Some of the protein amino acids have not yet been synthesized under prebiotic conditions. No one has devoted enough time to really synthesize all of them, but I think it can be done.

The real problem is the one you raise now: How were polymers synthesized? I report the attempts of Dr. Ibañez to synthesize oligonucleotides by using two methods: one that utilizes cyanimide as such — not dicyandiamide — and the other utilizing imidazole. The yields are not large.

All together we have synthesized up to roughly 1 per cent amounts of di-, tri-, and tetranucleotides, with very small traces of pentanucleotides. From 50 to 70 per cent of the di- and trinucleotides formed have the 3', 5' configuration corresponding to the biological polymers (Ibañez *et al.*, 1971 a, b, c).

I was pleased to persuade Dr. Ibañez to follow the research work of Dr. Khorana's group on the synthesis of oligodeoxynucleotides, and he was able to duplicate in my laboratory what they had done. After he mastered their technique, I asked him to use the cyanamide as a condensing agent and as a monomer he used the thymidine 5' monophosphate. I admit the mononucleotide would have to be synthesized first, but we started with the thymidine 5' monophosphate, because we had to start some place. I believe a very modest beginning has been made on the prebiotic synthesis of oligodeoxynucleotides. Comparable, not identical, results between these two compounds, cyanamide and imidazole, were obtained. I think more work must be done.

LOHRMANN: Is that in aqueous solution?

ORÓ: Yes.

MILLER: What sort of concentrations?

ORÓ: The concentrations are relatively low; I cannot recall. I don't have the papers here. I was not preparing to talk about this. I'll be very glad to send copies to all of you.

Leslie has probably seen a copy of this paper.

ORGEL: I think they were yields of about a per cent.

ORÓ: Maybe all nucleotides taken together are at the 1 per cent level. They decrease as you increase the degree of polymerization, the pentanucleotide is in very low yield, maybe 0.05 per cent (Ibanez *et al.*, 1971a). You may ask: Is this a good characterization? Since some of the preliminary reports came out we have done additional enzymatic work, and to our satisfaction from 50 to 70 per cent are 3'-5'-phosphodiesters.

Now, I would like to raise another point. I think some investigators suggested that perhaps ribonucleotides originated before other macromolecules. This is a circular type of reasoning. Many things went on simultaneously. I'm what geologists call uniformitarian, which in a biochemical context means: whatever happened in prebiological times in terms of conditions and major organic reaction pathways should not be significantly different from what is occurring today in living organisms. In this context, I eliminate high temperature polymer synthesis, not only because many liable compounds (e.g. sugars) do not survive the treatment, but mainly because the polymers must then be degraded in order to again start the synthesis and transfer of information processes under low temperature conditions.

I would prefer to use methods involving aqueous solutions, since life is a system containing about 70 per cent water. Second, DNA was first, not RNA. Aside from the uniformitarian principle, there are three additional lines of evidence in favor of this conclusion. (1) If there is ammonia, no matter how little, the environment is basic and in a basic environment DNA is stable, RNA is not. (2) The DNA has a built-in selectivity for the formation of $2'$-$3'$-phosphodiester linkages that the RNA does not have. (3) As a corollary of (2), DNA can form a linear polymer capable of simple duplication. This is more difficult for RNA and protein.

We have started with DNA-type mononucleotides, and it seems to work. Although we have yields less than one per cent, I think this is a beginning.

HULETT: I have never really found anybody who has actually working in the field that I disagreed strongly with. It bothers me that my daughter's high school textbook says this occurred, we know what happened, and really we don't know what happened.

Some people like to forget about the nucleotides completely at the origin of life, and talk simply about polypeptides, which somehow, in some way, then later developed. Is there anyone here who wants to defend that particular position?

SAGAN: There's a nice defense attempt in the book by Miller and Orgel (1972).

USHER: Eigen has already proved it can't be!

SAGAN: Most people seem to agree that it is not the case.

MILLER: Are you saying that we have to go all the way with DNA?

SAGAN: Yes, all the way with DNA! [Laughter] That's my rephrasing of a talk Stanley Miller gave on DNA chauvinism.

NAGYVARY: I am not aware of any particular important paper on the synthesis of deoxyribonucleotides. Would you suggest a plausible method toward the synthesis of deoxyribonucleotides?

ORGEL: Monomers or polymers?

NAGYVARY: From deoxyribose? I am aware of one or two reasonable theses on ribonucleotides but none on the synthesis of deoxyribonucleosides. If we don't understand formation of the monomer then it is hard to argue about the predominance of DNA polymers.

MILLER: Perhaps it hasn't been said clearly: there is no wholly satisfactory published synthesis of any nucleoside or nucleotide.

NAGYVARY: We have something in press.

MILLER: There's a very neat prebiotic synthesis of cytidine, but it gives primarily the alpha, instead of the beta configuration.

ORGEL: Does it still?

MILLER: This may have changed! This is a negative result that will probably change rapidly in the next few months. So far it has been very difficult to get any nucleoside in any decent yield.

NAGYVARY: I am aware of unpublished data concerning ribonucleosides, but know about no data concerning deoxyribonucleotides. This is a somewhat academic discussion at present.

PONNAMPERUMA: The nucleotide can be made from the nucleoside, but the difficulty is making the ribonucleoside.

ORGEL: There is a general point worth making. Polymers that occurred on the primitive earth may not have been as clear-cut as ours nowadays, because there weren't any enzymes then. I don't see why we should be obliged to choose either the deoxy- or the ribo-, or anything else. The original polymers may have been made from a rather heterogeneous collection of sugars, that included both the deoxy-, the ribo-, the arabino- and others.

NAGYVARY: A very neat compromise! The first polynucleotide contained an arabinose, not ribose, because these compounds are very resistant to hydrolysis. Arguments against the formation of ribonucleotides, instability at alkaline pH, and so on, would not apply to polyarabinonucleotides, because they are resistant to both acid and alkali. Particularly at 0° they may be stable for millions of years.

MILLER: Stability with respect to what?

NAGYVARY: To hydrolysis.

MILLER: Is the phosphodiester linkage stable too? Does the base hydrolyze off the sugar rapidly?

NAGYVARY: We are not aware of any means of removing the base from the sugar at present.

USHER: Do they not resemble DNA? Are they more stable?

NAGYVARY: Polyarabinonucleotides are much more stable. Their enormous stability would allow a gradual increase of chain lengths, leading us from monomer to polymer chemistry. The increase of chain lengths itself represents a decrease of entropy, and in a way, an evolution. Eventually the polyarabinonucleotides must be transferred to the polyribonucleotides, or reduced to a DNA-type compound.

USHER: I don't want to give an explanation that might be boring but Dr. Oró's objections concerning the stability of RNA are answerable if RNA is in the form of a double helix.

NAGYVARY: How is a double helix formed?

USHER: By starting with a length of template (it doesn't have to be very long) that then directs the formation of the complementary strand. Once this occurs the RNA double helix is stable, even though it is RNA, because of the geometry in the double helix, or even in the triple helix, or even in the polyadenylic acid double helix, which is actually parallel to double strands. Even in alkali the polymer would be stable in the 3', 5' linkage.

The other problem of having 2', 5' possibilities I think I know why that happens. I don't want to detail it, but it is not a problem either.

ORÓ: In these complex mixtures we probably have many things at the same time. Please don't misunderstand. I'm not trying to make an exclusive case for DNA, it is just that DNA is relatively more stable to basic conditions. The primitive environment was supposed to be somewhat alkaline. Secondly, RNA will tend to branch. There may be ways to overcome these difficulties. Certainly I am well aware of the work of Dr. Nagyvary. I'm not attempting to rule out his arabinonucleotides as an alternative to the problem. Our system is not complicated, so conceivably we can increase our yields.

USHER: What was your experiment? in one the cyanogen-containing compound was used and in the other imidazole?

ORÓ: Yes.

USHER: What's his imidazole got to do with it? I don't follow the chemistry.

ORÓ: Are you aware of the experiment of Pongs and Ts'o?

USHER: What was that?

ORÓ: It was published two years ago (Pongs and Ts'o, 1969). Their reactions were in a nonaqueous system. We just extrapolated from there to aqueous systems (Ibañez *et al.*, 1971).

USHER: In aqueous solution with imidazole you form 2',5' polymers?

MILLER: What is your source of energy?

ORÓ: This is a condensation reaction.

USHER: At what pH?

ORÓ: Slightly on the basic side, about 8 (actually 7.5).

USHER: A monester will be a dianion at that pH. I can't imagine it would work.

ORÓ: I will gladly send you the two possible mechanisms depending on the pH (Ibañez *et al.*, 1971 b, c).

NAGYVARY: Have you carbodiamide?

ORÓ: In this one there is no carbodiamide (imidazole).

ORGEL: People are puzzled about your source of free energy. There would be an equilibrium between the monomers and dimers, in which the dimer, would be much less than 1 per cent. The question is: where does the free energy come from? None of us can see a mechanism.

ORÓ: The same that you do in the other synthesis. Most of the syntheses are at temperatures slightly under 100°C.

ORGEL: Our experiments are either conducted with compounds which contain a preactivated material or they are conducted in the presence of a condensing agent. More recently (which we haven't mentioned) our experiments are done in the absence of water on a solid. I don't see how the condensation can occur in your case unless there is some trick you haven't told us about.

ORÓ: If that experiment is kept for 100 years, will you find any? This has not been done. We heated that one day, and then analyzed the product.

ORGEL: Have you heated it dry or in solution?

ORÓ: All in solution. If you have not read the paper I sent you, I'm sorry.

ORGEL: I read it. I couldn't understand it.

SAGAN: This may be a topic for cocktail hour!

Cyril, did you want to say something?

PONNAMPERUMA: Yes. We have analyzed the Murchison meteorite a Class II carbonaceous chondrite, and identified 18 amino acids, six of them are familiar and 12 are nonprotein (Kvenvolden *et al.*, 1970; Kvenvolden *et al.*, 1971).

We detected a similar crop of amino acids in the Murray meteorite, also a Class II carbonaceous chondrite (Lawless *et al.*, 1971). We recently were given a piece of the interior of the Murray meteorite. We have confirmed John Oró's recent work on both the Murchison and the Murray.

ORÓ: We are confirming your work.

SCHOPF: Do both of your results agree concerning the amount of racemization, distribution and relative concentrations of amino acids, in the Murchison?

PONNAMPERUMA: We do in the case of the Murchison. But I think the results on the Murray are different because we had an interior piece. John had a more contaminated piece. I think we were able to see the material a little more clearly in the case of the Murray.

My point is: we searched for nucleotide bases in both meteorites and we could only come up with 4-hydroxy pyrimidine. There was no adenine or guanine. Clair Folsome spent about three months on a very careful assay of this (Folsome *et al.*, 1971).

In these meteorites are some conditions which may be described as prebiotic, and amino acids are formed, in good yield: in the range of 10 microgram. However we cannot detect the bases. Even on the 5-hydroxy pyrimidine the evidence is still preliminary.

MILLER: Is that a pyrimidine ring with only one substitution on the 5 position?

PONNAMPERUMA: Yes. It is the only one that can be extracted. There may be some other substitutes; one difficulty is our lack of standards. Again we come back to requiring mass spectroscopy. We can only reasonably extrapolate (identify from mass spectra) with an authentic standard. We have tried every available purine and pyrimidine and we don't see anything.

MILLER: Is the basis of your identification the mass spectrum?

PONNAMPERUMA: Gas chromatography combined with mass spectrum.

MILLER: Without a standard?

PONNAMPERUMA: In this case with a standard. If a meteorite gives us genuine prebiotic conditions, they are conditions in which amino acids are synthesized, and not purines and pyrimidines. We know the synthesis of adenine is a commercial process from ammonium cyanide –

SAGAN: A few years ago didn't you say that as the hydrogen abundance varied the yields would go from preferential amino acid production to preferential nucleotide base production?

PONNAMPERUMA: No, we didn't conclude that. We found there was more adenine produced if there was less hydrogen.

It's very curious how this was discovered. Our experiment used Stanley Miller's mixture of 25 cm each of methane and ammonia, 100 ml of water, and 10 cm of hydrogen. One fine day we forgot to add the hydrogen and we found that far more adenine was synthesized! We then worked out the relationship, and found that with less hydrogen there was more adenine. We have no figures however for amino acid yields.

SAGAN: This is extremely interesting. As any planetary atmosphere evolves, it will have less and less hydrogen. If relatively more hydrogen is needed for amino acid production, then hydrogen depletion is correlated with the start of base production. This would be most interesting for the origins of life. It would also have interesting implications for the results of your meteorite studies: the parent body of Murchison was hydrogen rich, which is quite stunning, because that implies solar nebula origin.

PONNAMPERUMA: The Orgueil meteorite has also been studied by Anders and Hayatsu (Hayatsu, 1964). They claim to have found by chromatography adenine and guanine in it. That is a Class I carbonaceous chondrite – and they have found melanine and aniline is there.

We are now analyzing the Orgueil meteorite; hopefully next week we will know if there are bases present (Folsome et al, 1972).

SAGAN: The correlation I just suggested may be erroneous, but if there were such a correlation, then certain meteorites would have had to come from certain classes of parent bodies. That would be most important.

MILLER: In electron beam experiments, though they may have started out with zero or 10 cm of hydrogen, a good deal of hydrogen was produced during the irradiation. In this sort of experiment, one must be very careful to make sure the hydrogen pressure is constant before drawing sweeping conclusions.

SAGAN: Right.

PONNAMPERUMA: Much hydrogen is produced in an electric discharge experiment.

LEMMON: With a standard 45-minute irradiation, there would have been less hydrogen in regardless, though it was made. Stanley, did you ever forget to add hydrogen?

MILLER: Yes.

LEMMON: Was there an effect on the amino acid production?

MILLER: No, the results were just the same.

LEMMON: Apparently what you were asking has been done, Carl.

PONNAMPERUMA: In the case of adenine the results were clear.

SAGAN: If we believe what has just been said, it does appear that with an excess of hydrogen, amino acids are formed, but base is not; and as the hydrogen abundance decreases, amino acids are still formed, but nucleotide base production goes up.

MILLER: This sort of conclusion ought to rest on elaborate and carefully controlled experiments. If the experiments are not designed for this purpose, your conclusions may be erroneous.

SAGAN: A related question arises from discussions of the production of organic compounds in the atmosphere of Jupiter; namely, what happens with a huge excess of hydrogen? Suppose there is a thousand times more hydrogen than methane, ammonia, or water. What effect will this have on organic synthesis?

USHER: Will there by any phosphene?

SAGAN: Does one get any?

MILLER: Phosphene is thermodynamically unstable in the presence of water. 10,000 or 100,000 atmospheres is needed.

SAGAN: Does anyone know about the organic chemistry in the presence of a great excess of hydrogen?

MILLER: The yields might be less, but hydrogen cyanide still should be there.

LEMMON: I think he's asking for example, how stable are amino acids in an excess pressure of hydrogen?

MILLER: They are kinetically stable. They are thermodynamically unstable under any conditions.

LEMMON: It really hasn't been studied — it ought to be.

ORGEL: There may be a competition between two reactions: nitrogen atoms on methane and nitrogen atoms on hydrogen.

MILLER: Nitrogen atoms on hydrogen are not very efficient.

ORGEL: But with a thousandfold excess.

SAGAN: Free radicals are made under these conditions, but the most likely thing to collide with is hydrogen. Therefore wouldn't the original precursors be reformed? Wouldn't that result in lowering the yield?

MILLER: Yes. Ferris' study with the NH_2 radicals apparently reacting with H_2 may be an example of this (see previous discussion, pages 83-85).

ORÓ: Basically, there ought to be regeneration of the initial components: ammonia, methane, etc. If applied to Jupiter, a very cyclic system might possibly cause production of organic compounds whether there are electrical discharges or whatever energy source, and then regenerating starting materials with hydrogen in excess. Jupiter probably provides a very dynamic system, the right kind of system needed for an origin of life. In order to develop a biological system some pathway for the removal of the higher molecular weight compounds formed is needed — they need to go into an environment where they are less hydrogenated, if this is possible.

I think our lab has confirmed the very good meteorite work of Dr. Ponnamperuma. The Murchison meteorite is a very fortunate fall that occurred just a little over a year and a half ago. It seems to be loaded with a number of organic compounds, and the problem of contamination has been minimized.

The analysis of the amino acids by direct water extraction was kind of a surprise. Until then everybody was hydrolyzing the amino acids, for example, with hydrochloric acid, and then analyzing them. That meant the hydrolysis of the bulk of the meteorite. The fact these amino acids can be removed by just passing water through it, and subsequently hydrolyzed has brought a different dimension into these problems.

I want to relate the Murchison work to previous work in our laboratory, where the opposite method has been used. First a meteorite was extracted with methanol, benzene, and then water, and the residue then hydrolyzed. This was done purposely to remove any superficial contamination, but, paradoxical as it may seem, we were doing the opposite. We were removing the true indigenous amino acids and leaving the contaminating microorganisms, which possibly were not dissolved. Upon hydrolysis of the previously washed meteorite we obtained a distribution of amino acids in which at least 90 per cent on the average, were of the L-configuration, very similar to soil amino acids or amino acids of bacterial cell walls. They were essentially L-with roughly 10 per cent each of D-alanine and D-glutamic acid. These are two of the major amino acids in the cell wall.

LOHRMANN: I would like to ask Dr. Ponnamperuma, when he extracts these meteorites and finds guanine and adenine, do you also find something similar to the cyanide polymers —

PONNAMPERUMA: We didn't find any adenine or guanine.

SAGAN: He was quoting the work of Ed Anders (Univ. of Chicago, Department of Chemistry).

LOHRMANN: Did Anders group find cyanide polymers or other organics with high molecular weights?

PONNAMPERUMA: Juan, do you recall that paper? I think they mentioned melamine and aniline?

ORÓ: Two urea derivatives, but not cyanide polymers.

LOHRMANN: Was cyanide detected?

ORÓ: It was not measured.

SAGAN: Cyril, what is your opinion of Anders' claim that a Fischer-Tropsch synthetic process is responsible for many of the organics found in carbonaceous chondrites?

PONNAMPERUMA: Organics can be made that way.

SAGAN: True, but they claim that the Fischer-Tropsch synthesis is responsible for the carbonaceous chondrite organics, rather than the kind of synthesis that most people are talking about today.

PONNAMPERUMA: We can make amino acids that way and by electric discharge. They don't have enough information to decide between one or the other, in my opinion.

Juan, do you want to comment?

ORÓ: I shouldn't continue talking, but I must answer this, because we have really worked on it. We have done several experiments by open system and by closed system Fischer-Tropsch. We spent 3 or 4 years on experiments with the normal system used in industry, the open system. No isoprenoid hydrocarbons were formed (Gelpi *et al.*, 1970). So far no isoprenoid hydrocarbons (C_{15} to C_{20}) have been obtained by the closed system either.

With respect to the bases, adenine, guanine, and some pyrimidines — we have repeated the work in a closed system Fischer-Tropsch. It is such a highly modified Fischer-Tropsch system that maybe it should not be called Fischer-

Tropsch. In these experiments we almost confirm 100 per cent of their results. The purines and pyrimidines can be synthesized efficiently. Basically a good environment or good reaction conditions to synthesize the precursors of these things are provided — possibly HCN and cyanoacetylene and so forth.

PONNAMPERUMA: Were amino acids formed in your Fischer-Tropsch?

ORÓ: We did not specifically look for amino acids, as I recall. I can't say whether or not they are formed.

Formaldehyde and Cyanide

PONNAMPERUMA: Mr. Chairman, we still haven't answered Dr. Hulett's objection: How is enough HCN and formaldehyde formed. The data implies there will be not enough. This can be done in our flasks, but that has nothing to do with what happened 4 billion years ago.

SAGAN: Does someone have a comment on this?

HUBBARD: What is a realistic intensity of the UV source for considering your proposals that we should have a surface catalyzed synthesis or an aqueous phase?

SAGAN: We know the UV flux on the surface of the primitive earth.

HULETT: It is a reasonable source; but for how long can it operate?

HUBBARD: My rate would never satisfy the Belmont Manifesto, because I never would get that kind of a yield.

SAGAN: We haven't yet got a Belmont Manifesto.

SCHOPF: I suspect Jerry is referring to Hulett's statement on page 63: "The intensity of irradiation in the laboratory was many orders of magnitude greater than in nature, drastically altering the relative rates of synthesis and of nonphotochemical degradation of intermediates. The question is whether the laboratory syntheses are reasonable, since they use much larger energy fluxes."

MILLER: Let's look at the numbers, in the case of the cyanide production. "An attempt was made to estimate the highest possible cyanide production in the face of these uncertainties" — this gives as much as 10^{-6} moles per square centimeter per year. In the present ocean that implies, I think, 10^{-8}. If it then accumulates over 10,000 years —

HULETT: But how long will it last? And then there is the question of temperature.

MILLER: At a low temperature, namely $0°$ and pH 8, the half-life for hydrolysis is about 10,000 years, so the concentration can be built up by a factor of about 10,000.

If there are concentration mechanisms, it can be built up even more. Given enough time, consistent with the rate of hydrolysis, enough will be formed.

HULETT: I agree that one specific area to be investigated is the question of how low the temperature was, particularly since probably it was lower in the early days. Yes, it may be protected by low temperatures. Whether a concentration as high as 10^{-4} or 10^{-5} can be built up I don't know. Probably this was possible in areas where it was concentrated.

MILLER: Concerning aldehyde, cyanide, ammonia, to amino acids, there are no data, but my feeling is that these reactions will occur at fairly low concentrations. I once tried to calculate this. It might work at 10^{-5} molar with the appropriate amount of ammonia.

HULETT: I am much more suspicious of building up that kind of concentration of formaldehyde than of HCN. My personal predilection is toward getting HCN polymers, which then can be converted into amino acids, rather than Strecker synthesis, simply because the photolytic aldehyde rate of destruction is so high. Once it is in the ocean, you are much better off.

HUBBARD: I agree, but how easily will it go into the ocean?

HULETT: The generation should be at very high altitudes. On the other hand, if it is generated in concentrations Jerry suggested then maybe there is a better way of protecting it.

MILLER: Another way is to make the nitriles directly in the electric discharge, as Cyril did.

HULETT: This doesn't require formaldehyde.

MILLER: Yes, but once say, alanine nitrile gets into the mildly basic conditions of the ocean, it will dissociate to acetaldehyde, hydrogen cyanide and ammonia. But as Cyril demonstrated, at least the problem of reaching safety in the ocean can be avoided by simply making the nitriles directly.

BUHL: I would like to keep the door open to the possibility of some contribution from the interstellar medium of compounds like hydrogen cyanide and formaldehyde. We will discuss this on Friday in detail (see page 219).

HULETT: But consider the low concentrations. Even though we are sweeping up a fair amount of total HCN or formaldehyde, it is a very very small amount as compared to what can be generated by UV or electrical discharge.

SAGAN: Let us defer this discussion until Friday.

Are there other responses to Cyril's underlining of Dr. Hulett's quandry?

LEMMON: It still seems to me that with active surfaces on which to concentrate these monomers, coupled with the energy considerations we have discussed and a highly unsaturated HCN analog to perform the dehydration condensation — that we really have the mechanism to go to high molecular weight polymers, which, I admit, we have not found yet in the laboratory to any great degree. Oro's work is excellent. We would like to see far more than just four monomeric units. But in principle we have the mechanism at hand, conceptually.

HULETT: I'm perfectly willing to agree as long as you say "possible." When you start to say "almost inevitable," then you lose me!

SAGAN: Let me just say a word about the quantity of molecules produced. The numbers I have for our H_2S long wavelength synthesis are not many orders of magnitude different from those of other people.

Let us ignore all destruction of the amino acids we produce (Obviously, a highly optimistic assumption, which I will shortly retreat from). Dr. Bada will discuss this tomorrow.

Let us assume that the photon acceptor H_2S is there for some period of time, say, a billion years in optically thick quantities (parts per million). From models of solar evolution we know the ultraviolet flux. We ask: how much amino acid has been produced over the first billion years of earth history? It calculates out to be an astonishingly large number – about 100 kg per square centimeter column. This means, essentially, all or a little more than all, of the carbon in the sedimentary column has been processed through amino acids. Obviously amino acids are only one kind of organic compound which is made in this way.

Even if we lower that number by many orders of magnitude due to destruction, we still have a fantastically large concentration, 1 per cent solution of amino acids in the primitive ocean, doesn't seem to be completely out of the question. To get embarrassingly low concentrations, huge destruction rates are needed.

HULETT: In terms of molecules per square centimeter per second, the rates are huge. In terms of moles per square centimeter per year, they are very low. In terms of moles per square centimeter over a billion years, certainly they are huge.

Prebiotic Syntheses of Higher Order Complexity

HULETT: There is no question that life is on earth. What are the chances of life on any other solar system within a particular range of its sun-like the earth? My reaction is: probably very, very low. Somebody else's reaction might be: probably very, very high.

MILLER: Where is the most serious problem – in the formation of the monomers?

HULETT: No, synthesis of the polymers. Most of the monomers can certainly be made. We don't know exactly what the final concentrations are, but even starting off with unrealistic concentrations to make the polymers necessary, at the kind of rate necessary to have evolution going on over a long period of time before the first appearance of a photosynthetic mechanism, seems to be most difficult.

LEMMON: Think of how you would approach the problem of the monomers if this discussion were taking place 25 years ago. You would say: We will never be able to make these et cetera.

PONNAMPERUMA: Or go back to 1828, to Walter —

ORÓ: May we suggest that the next meeting, the last in this series, be held 10 years from now?* [Laughter]

MARGULIS: The goal obviously is not just the synthesis of polymers but coupling the amino acids with the nucleotide polymers. We might go on to discuss the data that Leslie has been mumbling about under his breath in this context. Has any progress been made?

PONNAMPERUMA: We're trying hard.

ORÓ: Not in our lab. I think Leslie's group probably has done more.

PONNAMPERUMA: We have done some. Carl Saxinger in my lab has tried to look at the relationship between the absorption of bases by a column to which he has attached some amino acids. We begin to see some kind of selectivity. There is what might be called a one-to-one reaction.

MARGULIS: Have you actually found nucleotide-peptide linkages?

PONNAMPERUMA: No. By covalent linkages we struck some amino acids onto an ion exchange column, and then poured a mixture of nucleotides through. Some ions respond preferentially. We have just begun this work.

ORGEL: Our experiments along these lines have all been total failures. We have a lot of excellent total failures. If you use amino acids with adenosine, for example, you find it doesn't make a bit of difference whether you use D-phenylalanine and L-phenylalanine.

MARGULIS: Why does it bother you?

ORGEL: Because if the claim is that the reactions are specific, they must depend on stereochimistry of the interactions. There is a greater stereochemical difference between D-phenylalanine and L-phenylalanine than say, between D-phenylalanine and glycine. Yet they react exactly the same way.

WOLMAN: At what temperature were these reactions run?

ORGEL: About 0°C.

---

*Ed. Note: Since the fifth meeting that was to take place in 1972 has been indefinitely cancelled from lack of funding, it is likely that even Dr. Oró's date is a bit optimistic on the early side.

SAGAN: If there is any category of amino acids which selectively interacts with polynucleotides, compared to other categories of amino acids, is not that already a step up in the origin of protein and the genetic code?

ORGEL: A good deal is known about this. Undoubtedly there are specificities. Probably the most clear-cut examples are changes of absorption spectra of polyadenylic acid on adding amino acids. It's clear that amino acids interact differently, but the results don't fit in very nicely with anything anyone knows.

Carrying this further, there doesn't seem to be any sensible relationship between the structure of the present genetic code and the particular amino acids which react best will particular nucleotide polymers.

SAGAN: Taking all the data at face value, I conclude there was some primitive stereochemistry which made a primitive code very different from the present one. The present code is the result of natural selection, whereas the primordial code was adequate for the origin of life. [See the discussion by Dr. Sagan in the first volume in this series Margulis, L. ed. 1970]

ORGEL: There are many discussions of this, but they are not very scientific.

NAGYVARY: Who made this experiment with polylysine, which can accommodate a certain number of adenylic acids?

ORGEL: Probably the best and earliest experiments were those of Felsenfeld, who showed the nucleic acid which reacted preferentially with polylycine, and a nucleic acid in which guanine reacts somewhat preferentially with polyarginine. That was the very first experiment of this type; probably ten years ago. There are many subsequent ones which culminate in some very good NMR studies of the amino acid-nucleotide interaction.

NAGYVARY: Someone in Alabama has shown that for every amino acid in polylysine three adenylates can be accommodated. It was published in *Nature* two years ago.

ORGEL: Much work has been done, undoubtedly there are specific interactions, but no one has been able to conclude very much from this.

SAGAN: The origin of the code is a worth-while subject for an entirely separate conference, particularly some years hence, when there is data. It is a very essential question, even if all monomers and all polymers can be made with acceptable ease. It is an important question, but not the answer to the origin of life.

I have been thinking about your daughter's textbook. It strikes me that there is a difference, looking at this subject from the inside and looking from the outside. We see all the specific problems in getting a particular compound, and so on, but there is the necessity of a broader view. Compare the point of view 40 or 50 years ago, with today. What is striking is that the sorts of monomers needed for life are very much the sorts of monomers that come out of the kinds of experiments we have heard about today. It is as if the origin of life (at least, in terms of the monomers) is built into the chemistry and physics — is in the cards. I suspect this quite striking correlation impresses writers of high school biology texts.

MARGULIS: You mean to say God, like the rest of the engineers on route 128, is slowly being put out of a job?

HULETT: I don't think writers of high school biology texts really know much about it. They read an article in the newspaper about a new discovery on origin of life, and their natural reaction is to think that it has been proven.

It appears to me that compounds present in life should be expected in the kinds of experiments you are talking about, because these compounds don't take a tremendous amount of energy to make. Otherwise life couldn't use them. You would expect them to be more easily produced than extremely unstable compounds. Compounds in living things have to be reasonably stable or nothing will last long enough to allow life to reproduce itself.

MILLER: Do you mean kinetically or thermodynamically stable?

HULETT: Kinetically.

MILLER: Some examples like cysteine are not very stable. On geological time scale cysteine is very unstable and so are free sugars.

PONNAMPERUMA: Adenine is very stable, on the other hand.

MILLER: Yes, but cytosine deaminates rapidly to uracil.

PONNAMPERUMA: We can live with uracil.

BADA: From your statement about possible life in other parts of the solar system, I gather that you feel a possible mechanism for the origin of life on earth is transportation of life from some other place.

Perhaps that happened, but we will need a source of carbon for these creatures to utilize. Maybe they are photosynthetic. Maybe they utilized the carbon dioxide in the atmosphere.

PONNAMPERUMA: Can we have one good reason for how that transportation took place?

SAGAN: Leslie Orgel and I have been arguing for hours about this.

BADA: Still organic material must have been synthesized on earth. Obviously we have sophisticated life now, so we must have had some early means of synthesizing organic material.

MILLER: Any organism other than one that's totally autotrophic —

ORGEL: Take a totally autotrophic organism.

HULETT: That would survive in the conditions of interstellar space.

SAGAN: May we defer the panspermia question to Friday? It is a hot and heavy issue.

SIEVER: From the outside looking in, it seems to me that nobody wants to rate the probability. Leslie talks about paradoxes, this is not a yes or no proposition. Dr. Hulett has been talking about probabilities. We are just talking about the probability of an occurrence. Who's estimate do you want to take, Harlow Shapley's or anybody else's. We are just reducing the probability that life occurred on a certain kind of planet, isn't that it?

HULETT: Everybody has his own innate feeling of probability, my innate feeling of probability of these occurrences seems lower than most of yours.

SAGAN: I'm convinced that this never going to be decided by argument. It has to be done experimentally. There are two kinds of methods in the fore. One is the search for organic matter and life on nearby planets That's NASA. The other is an attempt perhaps by radioastronomy to find evidence of highly organized organic chemistry on the planets of other stars, namely intelligent life.

LEMMON: Any group needs a devil's advocate. Maybe we have got a God's advocate here!

SAGAN: Can there be any pressing comments after that one?

ORGEL: Time for a drink!

— — — — — — — —

The meeting reconvened at nine-ten o'clock. Dr. Barghoorn opened the meeting.

BARGHOORN: About fifteen years ago there was an optimistic group (some of whom are represented here in this room) who entered into euphoria about what was called paleobiochemistry. We hoped we could trace biochemical systems back in time.

There has been considerable disillusionment in the ensuing fifteen years, partly due to the recognition of the geological contamination problem, independent of reagents, and the laboratory contamination problem. Another is the realization, through careful studies of the kinetics of organic compounds, of fundamental instabilities, with respect to long periods of time.

This morning our Discussion Leader, Jeff Bada, will be talking on limitations due to instability of organic compounds. I hope there still is some optimism about the stability of organic compounds. At any rate, I'd like to call on Jeff.

Amino Acid Racemization

BADA: I'm not prepared to discuss the broad problem of stability in all organic compounds that may have been present on the primitive earth. I want to mainly concentrate on amino acids, my major interest, and particularly I want to discuss some of the recent work I have been doing on the kinetics of racemization and the relevance of this to the problem of the origin of optical activity (something we haven't yet discussed).

The amino acids found in present day proteins are all of the L -configuration. Yet in all our prebiological experiments we synthesize a racemic mixture. What set of events occurred by which organisms started using only one enantiomer of the various amino acids? I have grouped the three different hypotheses which have been proposed to answer this question. (Two of them are sort of similar, but they have been expressed by various workers.) The first may be called spontaneous crystallization. The idea here is that a racemic mixture is synthesized under some localized ideal set of conditions and one enantiomer precipitates out, leaving the outher in solution.

SIEVER: Are you referring to Calvin's (Melvin Calvin, University of California) idea? The first molecule gets started —

BADA: Yes, in a way, although Calvin didn't express it as a spontaneous crystallization.

Another process which has been used to explain the origin of optical

124

activity is an asymmetric synthesis, or asymmetric decomposition where by a special type of synthesis only one enantiomer is synthesized or by a special type of decomposition only one enantiomer is decomposed, leaving — or producing — just the one enantiomer, and the enantiomer is then incorporated into the first living organisms.

SCHOPF: Jeff, how much data are there that clearly establish in all abiotic syntheses racemic mixtures are produced? There are good reasons for believing this, but I'm just wondering how many experiments have been done. My impression is that maybe two measurements have been made.

BADA: If you start with nonoptically active reagents, such as cyanide, methane, and formaldehyde, it's impossible to produce optically active compounds.

ORGEL: If you say that the probability of producing one enantiomer is the same as the probability of producing another enantiomer, it is a very different statement.

BADA: Perhaps yesterday we should have included in our list (p. 34) the search for optical enantiomers in prebiological syntheses.

PONNAMPERUMA: To answer Bill: In an electric discharge experiment, we found a racemic mixture: aspartic acid, glutamic acid, and alanine. It was only one experiment.

BADA: Stanley, didn't you also find a racemic mixture?

MILLER: Yes. In the original experiments I put alanine in the polarimeter, and it was zero in rotation within experimental error.

PONNAMPERUMA: Syntheses on certain surfaces might be interesting. Claims have been made that amino acids have been synthesized from methane, ammonia and water, have been passed over heated sand — what is the optical activity of sand?

MILLER: Quartz is an optically asymmetric crystal.

SIEVER: Left or right?

MILLER: Both. One count of 5,000 quartz crystals in the same quartz bed was, within the statistics, fifty per cent right and fifty per cent left.

SIEVER: In one survey it turned out that 50.4 per cent were lefthanded, which assures me that the universe is lefthanded!

The statistics weren't very good. He simply counted all quartz crystals he could get ahold of.

MILLER: But 50.4 per cent is closer to the 50 : 50 ratio than statistics would predict.

BADA: Even if one enantiomer can be synthesized or in some way produced, in order for this to be preserved it must be prevented from racemizing. I have studied the kinetics of racemization of several amino acids, as a function of pH, and temperature, to get some idea of what the half-life of these racemization reactions are.

First we'd like to know pH dependence of the racemization reaction. In Fig. 8 I have plotted the racemization rate versus pH. These are data at 118°C for aspartic acid.

USHER: How do you regulate the pH at 118°?

BADA: A buffered solution of phosphates, borate, etc. is made: the National Bureau of Standards has standard values for these buffers to 95°.

MILLER: They are very good.

BADA: We determined the ratio using a polarimeter. They are heated in ampoules sealed under vacuum for various lengths of time, and the change in rotation is measured on a polarimeter.

The kinetics are quite complicated. They can be interpreted in detail but I won't do that now because it is not relevant. What is of interest is that in the pH range that interests us (between pH 7 and 9), there is very little dependence on pH at least for aspartic acid. Aspartic acid is a dicarboxylic amino acid. Other alkyl or aromatic substituted amino acids have a different rate vs pH profiles, but in the region pH 5-8 they are also independent of pH.

Therefore, the effect of pH on racemization kinetics is really not too great in the region around neutral pH. The argument that the pH of the ocean was such that one enantiomer might have been stabilized by raising or lowering the pH is not important.

SOFFEN: Why are the curves that way? Is it because of the carboxyl?

BADA: It is related to aspartic acid in various ionic forms, depending upon the pH. It is due to the racemization of the various ionic species. The whole mechanism can be written out and it is very complicated, but these kinetics can be beautifully predicted.

In Fig. 9 I have plotted the rates of racemization of several amino acids at pH 7.6 as a function of temperature. The quantity k is the rate constant for interconversion of the D and L enantiomers and 2k is therefore the rate of racemization. The relative rates of amino acids can be seen, more or less. Isoleucine is the slowest to racemize. (We actually know isoleucine does not racemize, it epimerizes. I'm using "racemization" as a general expression for change in optical configuration only at the α-carbon.)

Valine, although not plotted here, is about the same as isoleucine. Alanine is a little faster. Phenylalanine falls between alanine and aspartic acid, but has a different slope. Phenylalanine has a different slope because of the pK of phenylalanine and the ionic form that racemizes at that pH. Finally aspartic acid is fastest.

The half-lives we get from this are important.

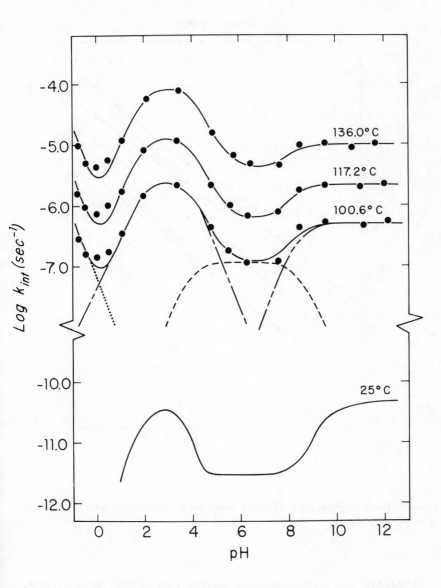

Figure 8. Interconversion rate of the D and L enantiomers of aspartic acid at 118°C as a function of pH.

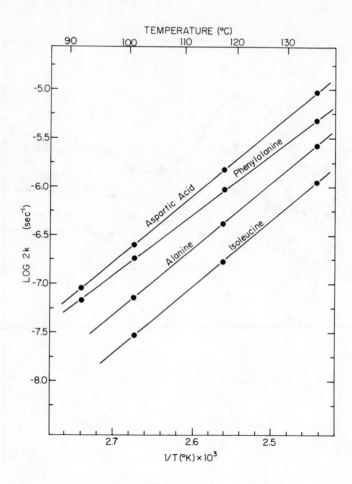

Figure 9. Interconversion rates of D and L enantiomers of four amino acids as a function of temperature at pH 7.6.

SAGAN: Jeff, qualitatively how do these racemization Arrhenius plots compare with thermal degradation Arrhenius plots?

BADA: They all racemize faster than they decompose.

SAGAN: But relative to each other, do those which decompose fastest racemize fastest?

ABELSON: There is no relationship.

BADA: It is a different mechanism.

SAGAN: Is there no statistical correlation between the two phenomena?

MILLER: Some, but it is not very clear.

BADA: Decomposition involves decarboxylation, deamination, dehydration, and so forth — these all can't be lumped together.

SANCHEZ: How does histidine fit into your plot (Fig. 9)?

BADA: I haven't studied histidine, sorry.

I was interested in this particular set of amino acids, because of my work in deep sea sediments. Histidine is not found in deep sea sediments, apparently it is oxidized very fast.

I have extrapolated these values down to 0° and 25°C (Table 6). At 0° phenylalanine is the fastest amino acid to racemize that I have studied so far, while isoleucine is the slowest. The racemization rate for isoleucine should be representative of alkyl substituted amino acids; aspartic acid of acidic amino acids; and phenylalanine of aromatic substituted amino acids. The only group I haven't studied is hydroxy substituted amino acids, serine and threonine. These decompose quite rapidly. Rates of decomposition of serine and threonine are at least comparable to the rates of racemization. The rates of racemization are all much faster than the decomposition rates for the other amino acids.

TABLE 6. HALF-LIVES (YEARS) FOR RACEMIZATION OF AMINO ACIDS AT pH 7.6 AND IONIC STRENGTH OF 0.5.

	0°C	25°C
Phenylalanine	1.6×10^5	2.03×10^3
Aspartic acid	4.2×10^5	3.46×10^3
Alanine	1.1×10^6	1.10×10^4
Isoleucine[a]	4.4×10^6	3.47×10^4

[a]These would be the half-life values if the equilibrium ratio of alloisoleucine to isoleucine was 1.0. At 140° the ratio is 1.25. The temperature variation of the ratio is not, however; a value of 1.0 is therefore assumed for 0° and 25°C. The uncertainty arising from this assumption is probably less than 10%.

As shown in Table 6, the racemization half-lives are fairly short, ranging from a few hundred thousand to five million years. However, some people might argue that five million years is long enough. A stereoselective decomposition occurred, resulting in the formation of one enantiomer, and then life arose. The values in Table 6 are probably maximum estimates. The actual half-lives for racemization in the primitive ocean are probably even shorter, because amino acids can be chelated by various metal ions.

It is difficult to estimate the concentration of metal ions like the cupric, manganese, or ferric ion, in the primitive ocean. However, what is important is the amount of amino acids which would be complexed by metal ions in the primitive ocean. This calculation involves using the equilibrium for the metal ion-amino acid complex at the pH of the ocean. I have carried out these calculations for cupric ion complexes in the recent ocean. From these calculations and the racemization rate of alanine in some cupric metal ion adentylates, I have estimated the metal ion catalyzed racemization rate in the present ocean (Bada, 1971). These calculations give a racemization rate for chelated alanine which is almost a factor of 100 greater than that for nonchelated alanine. The half lives in Table 6 are therefore maximum estimates of the racemization rates in the primitive ocean. The actual racemization rates were probably substantially higher because of catalysis by chelation of amino acids by metal ions.

These values imply amino acids would be rapidly racemized in the primitive ocean, even if they were synthesized by some stereoselective method, or produced by some stereoselective decomposition.

SAGAN: What is the implication for your chelation calculation?

BADA: It is to give you some idea of the magnitude of the effect chelation has on the racemization rates of the amino acids.

SAGAN: The real situation is worse.

BADA: Yes. The half lives in Table 6 are maximum values. The metal ion catalyzed reaction is $\sim$ 100 times faster and these would probably give minimum half lives for racemization. The actual rate is probably some place in between.

USHER: These are extraordinarily temperature sensitive. I'm not sure but I expect low temperatures would be actually safer. Do you have any evidence for that?

BADA: Even at $0°C$ the rates are substantial and I don't believe the temperature of the primitive ocean was lower than this.

USHER: Did you study the whole pH range at other temperatures too?

BADA: Oh, yes, at $100°$, $118°$, and $135°C$.

The racemization rates for the various amino acids show that any asymmetric synthesis or decomposition will not solve the problem of the origin of optical activity. Since the problem is not solved by these mechanisms, I can just list a third alternative: that the use of one enantiomer by an organism has an advantage to that organism.

I can't say this is definitely how optically active compounds became incorporated into the first organisms, but only that once organisms began the process of protein synthesis, etc., it would have been to their advantage to use only one enantiomer and it was by chance that only L-amino acids are used by terrestrial organisms. If life exists in other parts of the universe, we would expect the numbers of planets on which organisms use only one amino acid enantiomer to equal the number which use the other enantiomer.

SAGAN: It is like giving orders to a carpenter to build a house with random mixtures of right- and left-handed threaded screws. Obviously it is much easier to give him one form or the other, and it doesn't matter which one.

BADA: This discussion also suggests that the origin of optical activity may not have taken place until life had progressed to a more advanced stage. Perhaps protein synthesis was fairly efficient and at this point it was an advantage for an organism to ultilize only one enantiomer.

USHER: Do we accept Eigen's stoichiastic theory calculations? Because if we do, this really is no problem. Since it hasn't been published, let's arrange for a copy to be sent to us. By nonequilibrium thermodynamics, and heaven knows what else, he calculates that it is a frozen accident: one "reproducing" system began which necessarily took over what he calls the growth phenomenon.

MILLER: There are various kinds. There is the frozen accident that took place early in the development of the first replicating organisms; and there is the frozen accident that occurred in the amino acids that formed in the ocean prior to the formation of those organisms.

USHER: He doesn't actually mean the first organisms. The frozen accident in his calculation occurs at the onset of his growth cycles – interactions of nucleotides and amino acids – that shows an increase in information. It is selective at that level.

SAGAN: At the next meeting of this group we should invite Eigen who claims to have an inductive theory of natural selection and the origin of life.

USHER: I second that, Carl, except that we will never be able to talk ourselves!

LEMMON: Something in favor of the biological development of L-amino acids and against the frozen accident idea, is that it is not quite true that our proteins only have L-amino acids. A number of bacteria, procaryotic cells, have up to ten per cent of D-amino acids in their proteins.

ORGEL: No, not in their proteins, in their cell walls.

MILLER: They are all L-amino acids in their enzyme proteins.

LEMMON: In proteins, I don't know which.

ORGEL: They are not proteins, they are in peptides in cells walls.

ORÓ: Mucopeptides.

ABELSON: It is curious about isoleucine that in this change from

L-isoleucine to an equilibrium with D-Alloisoleucine the number of molecules is not equal. There is about 1.3 times as much alloisoleucine formed as in the equilibrium mixture, as isoleucine of the L-isoleucine.

There is also the corresponding pair of D-isoleucine and L-alloisoleucine. I don't know what that equilibrium mixture is.

Then there is the equilibrium involving threonine, a comparable situation. I don't know what the equilibrium mixture is there, but it turns out 1.3 times as much allo- in this example for reasons known to God and physical chemistry and the way atoms nestle together. Possibly with one of these amino acids, for example, threonine, the L-form actually is more abundant, and hence more favored.

BADA: L-threonine is more stable.

ORGEL: Your argument may be persuasive – to explain why we find isoleucine rather than alloisoleucine, but it can't explain why L-isoleucine is preferred to D-isoleucine.

ABELSON: In this example it is the wrong way.

ORGEL: Then it's impossible.

ABELSON: Perhaps in one of these other possible combinations it swings the other way.

ORGEL: Very interesting. Why is the less abundant form in proteins Stan?

MILLER: This is only about 30 per cent difference. That's peanuts compared to the optical purity needed.

ABELSON: If it went the other way around, it would explain everything!

BADA: It's not clear what the equilibrium constant is at low temperatures.

ABELSON: It is the same at low temperatures.

BADA: I have done experiments on the rate of racemization –

ABELSON: But we have measured it in sediments, in shells.

BADA: You may have measured the ratio of the decomposition rates, and not the ratio of the enantiomers of equilibrium.

ABELSON: We will disagree on that.

BADA: The hydroxyproline-allohydroxyproline ratio is 1.0 at 0°. There is no data on threonine or allothreonine.

ABELSON: They are difficult to resolve.

BADA: I'd like next to discuss my work on deep sea sediments. These studies give racemization rates we observe in the geological environment, and can be used to make estimates of whether the amino acids in ancient shales are indeed indigenous.

As was shown in Table 6 amino acids have racemization half-lives on the order of tens of thousands to millions of years. I therefore thought perhaps in deep sea sediment which cover time scales (depending on the source of the core) of several hundreds of thousands to several millions years that this racemization reaction could be detected.

We collected a 5 m core from the Atlantic fracture zone, in the middle of the Mid-Atlantic ridge and analyzed various sections from it. We studied the racemization of isoleucine since the determinations could be carried out on an automatic amino acid analyzer. We measured the ratio of isoleucine to alloisoleucine as a function of depth in the core.

Using these data, we plotted log (1+ alloisoleucine/isoleucine), which is the rate expression for irreversible first order reaction. Of course the racemization is a reversible first order reaction, but since we observed very little racemization in this particular core, we neglected the back reaction and only considered the formation of alloisoleucine from isoleucine. We plotted the above function *vs* depth, and got a very nice straight line.

The depth in a sedimentary column is directly proportional to time. Knowing the rate of racemization, we thought we could calculate a sedimentation rate or an age for this particular sediment.

The first data we used was our aqueous solution racemization kinetics. Based on the pH 7.6 data at $2°C$, the racemization rate of isoleucine is 1.2 x 10^{-7}/yr. Using this value we obtained a sedimentation rate of 4.2 mm per thousand years. The bottom of this core at 518 cm is 1.23 million years of age.

At that time we had no other chemical check on the age of this core. There are no good chemical methods to determine the ages of deep sea sediments in excess of a few hundred thousand years. There is a physical method, using paleomagnetics, which involves studying the orientation of the magnetic vector of the earth. We know the earth's magnetic field has fluctuated throughout the past, and at what times these fluctuations occurred. From these data we can calculate an age.

Unfortunately the paleomagnetics on this core did not give very clear-cut results.

So instead of trying to do another chemical method, like carbon[14] or some other radionuclide decay series, we decided to measure the rate of racemization in the sediment.

I first thought, intuitively, that the rate of racemization in the sediment would be slower than in aqueous solution.

The data on Fig. 10 are the rates we have obtained for the reaction in a deep-sea calcareous core. The values of k_{iso} which are shown are the rate constant for conversion of isoleucine to alloisoleucine; the dashed lines are the rates in aqueous solution at pH 7.6. Note that the rate of racemization in the sediment is faster than in aqueous solution at lower temperatures. We have also investigated the racemization of isoleucine at elevated temperatures in a core from Lake Ontario and these studies give rates nearly the same as those for the deep-sea sediment. The sedimentary environment therefore has no large effect on the racemization rate. We have proposed a mechanism to account for the difference in rates of racemization between the sediment and in aqueous solution, but it is complicated and not important to our discussion here.

ORGEL: What are amino acids like in sediments?

BADA: The amino acids exist as proteins, peptides and "free" amino acids in the sediment. The amount in the form of protein decreases with time because of hydrolysis.

USHER: What are you actually measuring when you say rate of racemization? Are you measuring a change in optical activity?

BADA: Oh, no! Not in the geological samples. This is from the amino acid analyzer: the amino acids are separated on the Dowex-50, and put on the amino acid analyzer and alloisoleucine is separated from isoleucine.

USHER: I see. And is this done for the first five per cent of the reaction?

BADA: No. These are samples of mud heated for various lengths of time to these various temperatures.

USHER: If you are measuring an approach to equilibrium, then the constant is the sum of the —

BADA: But we are not even close to equilibrium.

USHER: Yes, you measure about the first five per cent, or something.

BADA: 10.

USHER: 10 per cent. Fine! Okay.

ABELSON: Somebody asked about the form of amino acids in mud. Microorganisms love free amino acids. So in this mud the amino acid you ultimately get out can not be in the form of free amino acids.

BADA: I'm not so sure. You may be right about lake sediments but in the deep ocean, at least in the 5 m core we studied, microorganisms were found only in the upper 4 or 5 cm. Below that —

ABELSON: Do you think the organic matter was laid down under sterile conditions?

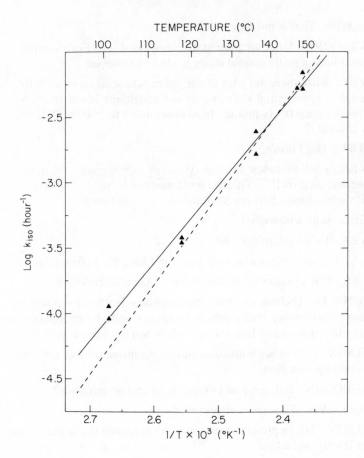

Figure 10. Rates for conversion of isoleucine to alloisoleucine. Solid line: sediment (deep-sea calcareous core) Dashed line: aqueous solution at pH 7.6.

BADA: No. A source of amino acids in deep sea sediments probably is organisms residing in the upper 10 cm. Often they die and are buried –

ABELSON: Aren't friends around to scavenge after they die?

SIEVER: It's got to end up some place.

BADA: There are no microorganisms below a depth of 10 cm in deep sea sediments.

ABELSON: That is not true.

BARGHOORN: What are your statements based on? For example has there ever been a really careful study made of anaerobes?

SIEVER: Rittenberg did a lot of careful experiments most recently. He sampled the experimental Mohole core at Guadeloupe Island and cultured microbes for up to six months from cores down to 500 ft. Bill, do you know this data?

SCHOPF: No, I don't.

SIEVER: I believe below the top 10 cm, or approximately that, he simply observed no growth, no matter what medium he used. We have discussed this various times, perhaps he didn't use the right media.

ORGEL: Is he a biologist?

SIEVER: He's a microbiologist.

BADA: It is always possible that you don't have the right medium.

ORGEL: It is a matter of aerobic against anaerobic conditions.

SCHOPF: Dr. Abelson, can free amino acids be leached from soils in various ways? Presumably they ought to be eaten readily by microorganisms but certainly the presence of free amino acids in soil has been reported.

ABELSON: Amino acids do come out of the humic acid of soil, but they are bound up somehow.

BARGHOORN: Bill, are you talking about free amino acids?

SCHOPF: Yes. This has been reported.

ABELSON: Try to preserve free amino acids against the action of bacteria. Give it a try some time.

SIEVER: No matter at what stage the process is interrupted, even if there are many bacteria, sooner or later at least one bacterium is going to die, lyze and produce some amino acids. To get rid of all amino acids you must inorganically oxidize all of the carbon, and any residue is obviously of biological origin.

ABELSON: In order to analyze amino acids on an amino acid analyzer, you need much more than one live bacterium.

ORGEL: This method clearly wouldn't work on live bacteria, because you would see not material at the age at which it was sedimented, but the last time it passed through a live bacterium. So you can't have it both ways.

PONNAMPERUMA: Even in a live bacterium hydrolysis is required to release the free amino acids.

ABELSON: I think it is unlikely that the amino acids are in free form; I think they are either in carbonates, or tied up in humic acid material.

BADA: I completely agree. They are not just floating around as ion exchangeable ions on clay or in the interstitial waters of these sediments, because the diffusion coefficient in these sediments is fast enough that if there were a racemic mixture at 5 m, the diffusion would wipe this out very quickly. If the amino acids were free to diffuse up and down the sediment, you would not expect to see a relationship.

ORGEL: How do you get them out?

BADA: We hydrolyze the sediment but in certain very rich carbonate sediments the carbonate can be dissolved in dilute acid and free amino acids come out.

Persistence of Peptides and L-Amino Acids in Sediment

This brings up the question of how long peptides persist. Our data aren't very clear.

MILLER: Have you data on the peptide hydrolysis?

BADA: Yes. I have just two experiments which I hesitate to discuss because it is hard to interpret. I also have some literature.

MILLER: Yes. I think we both calculated that peptides hydrolyze rapidly in aqueous solution, with a half-life of about 100,000 years, the rate depending, of course, on the amino acids in the peptide.

SAGAN: At what temperature?

MILLER: $0°$. This is important. – it implies prebiotic peptide synthesizing schemes have to operate continuously, because the peptides degrade at this rate. We know how long it took to make life, but presumably it was relatively short, compared –

BADA: Right.

LEMMON: Is this altered when peptides are adsorbed, stuck on a reactive surface?

MILLER: There are no data. It might increase or decrease the time. This could never be predicted. The data on which I base the 100,000 years estimate is very poor. It wasn't designed to get these numbers at pH 8. I suspect the true rate is considerably faster.

BADA: Yes.

LEMMON: The 10^5 years was at pH 7.

BADA: At pH 7.6 in peptide hydrolysis apparently the N-terminal ends are chopped away. The length of the peptide then determines the rate at which all the amino acids will be released.

ABELSON: No, it is more complicated. The combination in which valine is next to another amino acid is very difficult to break down.

BADA: Yes. I just meant a general statement. Obviously the R substituent has a big effect.

SCHOPF: Does this imply that polypeptides reported from Tertiary sediments and even from Paleozoic sediments can not be syngenetic?

BADA: I don't know these publications. I know people who have looked with no success.

SCHOPF: They are from Ordovician graptolites. But how about the peptides from the pelecypods? Would you agree that the geologic evidence is that there cannot be peptides any older than about 100,000 years?

ABELSON: Yes, that order. A few peptide links may go back to a million years, but not much more.

MARGULIS: But they claimed to find detailed *in situ* correlation with the morphology of the nautilus shells.

BARGHOORN: This is Gregoire and Florkin's work.

SCHOPF: Yes, Florkin, Gregoire, Bricteux-Gregoire and Schoffeniels, (1961).

ABELSON: They are basing the evidence on that crazy test.

MILLER: The biuret test?

ABELSON: Biuret!

SCHOPF: It is my understanding that pronase has been used in some studies, suggesting the presence of peptide bonds.

MILLER: Pronase or proteinase?

SCHOPF: It was an enzyme that would break a peptide bond.

MILLER: It sounds like pronase which self-digests! [Laughter]

PONNAMPERUMA: Pronase is very bad for that.

MILLER: If the data on the peptide hydrolysis is based on aqueous solution, and it's possible that when it gets in the shell it's inhibited in some way, but I doubt it.

ABELSON: No, it isn't.

BADA: It is probably hydroxide ion that performs the hydrolysis. It is hydroxide ion (or some other base) that's racemizing the amino acids too.

BARGHOORN: What are the temperatures and the pH of your Lake Ontario core?

BADA: The temperature there is 3°. The pH of the interstitial water is around 6.5. We are very excited about the possibility of using racemization kinetics to calculate paleotemperatures. It is no more than 13,000 years old. It can be dated very nicely by carbon[14]; we hope to then use the carbon[14] ages in our racemization kinetics to calculate the temperature in the core.

ORGEL: You must be very careful with that. If the racemization is a noncatalyzed reaction, then it is okay. But if it is catalyzed reaction, different sediments will come down in different patterns, and you will get different plots for each.

BADA: It doesn't seem to be so, though. The Ontario core is less than one per cent calcium carbonate. The deep sea core we studied was 75 per cent calcium carbonate, yet the rates are the same?

BARGHOORN: What are your carbon[14] dates?

BADA: The bottom of the core is 24,000 years old, based on analyses carried out by the carbon[14] laboratory at University of California at San Diego.

BARGHOORN: But there are Devonian spores in the Lake Ontario sediments. A small amount, but important.

BADA: I didn't know that. The pollen work is being done by the Canadians.

BARGHOORN: But the dead carbon might affect your dates.

SIEVER: How much would there be?

BARGHOORN: It would be a small amount, but it wouldn't take very much.

SIEVER: I'd estimate one part in about 10,000.

BARGHOORN: It might be a lot more.

SIEVER: Really?

BADA: I would like to discuss the probability of L-amino acids persisting for long times in geological samples, particularly the old shales, like the Green River shale and the Fig Tree cherts.

All these ancient cherts were originally deep sea deposits, or some type of sedimentary deposits. The racemization studies in sediments suggest that in the first few million years after a sediment was deposited the amino acids were racemized. After twenty-five or fifty million years, or at what ever time chert begins to form, what is incorporated into the chert should be racemic mixtures of amino acids. The observation by Dr. Abelson and others that only L-amino acids, or a predominance of L-amino acids, exists in the old cherts, is very strong evidence that the amino acids are not indigenous to the sample. They may have been introduced millions of years ago, but they are not the same age as the sediment.

SCHOPF: There's a difference between indigenous and syngenetic.

Presumably you mean they are not syngenetic with the original deposition; but clearly they are indigenous to the rock, as received in the laboratory.

BADA: Okay. Yes.

SCHOPF: At least it is not laboratory contamination.

BADA: My definition of "indigenous" is apparently different from yours. I expect that amino acids in very old geological samples will be racemized. Probably what has happened is that through groundwater flushing, or something, the original amino acids were removed; superimposed upon the original amino acids were the L-amino acids from recent organisms.

SCHOPF: Are there any reports of analyses of amino acids in sediments older than the Tertiary in which racemic amino acids have been found?

BADA: Yes. In very deep Deep Sea Drilling Project (DSDP) cores, brought up from 200 m or more, the amino acids are racemic.

BARGHOORN: What's the situation with Green River?

BADA: Cyril probably knows better, but I think Kvenvolden has found that there is only about ten per cent D-alanine in the Green River shale.

PONNAMPERUMA: It is very ambiguous. In some cases they appear to be completely racemized. In some there was a little more L- than D-. More careful work is needed, but most appear to be racemized.

ABELSON: Wherever the situation looks clean, there's no question about it being racemized. You find unstable amino acids in a Cretaceous sample and then it is not quite racemic — but wherever you find just amino acids that are stable and no serine in it, it's a racemic mixture.

SCHOPF: Have these analyses actually been published? They ought to be racemic, but has this actually been observed?

ABELSON: Our work on racemization of amino acid has been discussed extensively in Annual Reports of the Carnegie Institution of Washington.

BADA: The obvious next question concerns Cyril's and John Oró's work on the meteorites. Since you do find a D-, L-mixture of amino acids in certain meteorites, can it be argued that the amino acids were introduced during the fall to the earth and during the fall the temperatures became high enough to racemize these amino acids.

Our kinetics show that heating at 150° for days and weeks are required to racemize most amino acids. I therefore don't think that any contamination followed by racemization could have occurred in the meteorite samples. The possibility of this can be neglected.

MILLER: But if the meteorite was formed four and a half billion years ago, with organisms that made D-amino acids? Is it possible that they would racemize floating around in space?

BADA: I don't know the history of the Class II carbonaceous chondrites, how they were formed.

PONNAMPERUMA: There's no way of disproving that.

BADA: There's no way of disproving it! You may say that we're seeing the racemization product of some organism using D-amino acids.

MILLER: I think you can say more. You can go by the mechanism of the aqueous solution or sediment racemization. This involves the loss of the α-hydrogen to form a carbanion, and presumably that meteorite is pretty dry and has been outgassed very well.

BADA: It is difficult to say what is pulling off the alpha-hydrogen. In a calcium carbonate matrix it may be carbonate ion, water may not be limiting. It may be just the availability of hydroxide or some other base. In clays it could be a hydroxyl substituent on a clay or something.

MILLER: You mean it is possible that the Murchison amino acids four billion years ago were either D- or L-, and they have racemized since then.

BADA: Yes.

BARGHOORN: Here is a slight paradox. If these racemic mixtures in older rocks are the result of secondary displacement, free amino acids must be floating around in the groundwater. Abelson has argued against this.

BADA: There are free amino acids in groundwaters.

ORGEL: The levels are very low.

BADA: Our own work has shown amino acids in groundwaters but they are extremely low.

ORGEL: Bacteria pump in amino acids using membrane-bound enzymes. From studies of real bacteria we know they can only work at concentrations of about 10^{-5} moles or higher.

Bacteria therefore will not eliminate all amino acids from groundwater or any other water, once the concentration falls substantially below the kM in the Michaelis concept. Bacteria can not live on solutions which are too dilute, and for any given type of bacteria there is usually a lower limit with which they can cope.

BARGHOORN: The possibility that these rocks are contaminated directly by microorganisms is eliminated because microorganisms can not penetrate into them.

BADA: They may be on the surface, though.

ORGEL: Yes. No matter what the number of bacteria, there will still be very low amino acid levels. Very sensitive analytical techniques might still pick them up. Jeff, you can see 10^{-5} molar, can't you?

BADA: Yes.

ABELSON: Microorganisms vary a great deal in their avidity for individual amino acids. With *Escherichia coli* glutamic acid may be in the medium, but valine won't be around.

ORGEL: If the rate of growth of bacteria is plotted against decreasing concentration of amino acids in the medium, at first it would go down linearly, and then growth would stop. The bacteria stop growing, unless they synthesize valine for themselves.

BADA: That is exactly the argument why the sea at depths of greater than about 500 m is more or less sterile. Apparently the quantity of important organic nutrients, like amino acids and carbohydrates are just too low, below the threshold concentration for heterotrophic bacteria to utilize.

LEMMON: In your work where apparently adsorption of the amino acids on sediments made them more susceptible to racemization were they not something less than monomolecular layers on the sediment? I'm thinking you can't compare it in any way with a crystalline amino acid, which doesn't racemize this way.

BADA: No. Remember, I do not put amino acids into these sediments. We took sedimentary material which we know contained amino acids that are all L- or perhaps up to three percent D-.

LEMMON: Yes, and warmed them up.

BADA: These are amino acids routinely found in sediment that we heated at high temperatures to get the rate. I don't know now how they distributed in clay.

SIEVER: Referring to Phil Abelson's comments on the viability of bacteria at depth in groundwater this may be one of the most sensitive ways to find out, but another way is by isotope fractionation effects. From some of the JOIDES work by Kaplan's group (pers. comm) the inference from sulfur isotope fractionation is that fractionation by bacterial sulfate reducers continues at depth. At least, one can only account for those sulfur isotope data on that basis, since no other process involving that fractionation is known. It would be very useful to use this, coupled with isotopic fractionation, to find out.

USHER: If an L- or D- preponderance, there at the beginning, racemizes slowly over the entire period of time, is it possible that one of the isomers will show isotope fractionation, whereas a racemic mixture will give the same isotope distribution over an equal period of time?

MILLER: The effect on the carbon would be small; it would be better to look at the deuterium isotope fractionation rather than the carbon.

ORGEL: That should be very good. You should see a substantial effect.

BADA: How much deuterium is needed for that?

USHER: What is the percentage of deuterium in natural hydrogen?

MILLER: That's not the problem. It's the sample size.

USHER: How much sample would be needed to have enough material?

MILLER: You ought to speak to Harmon Craig (Scripps Institute of Oceanography, La Jolla, California) about that.

BADA: I have talked to Harmon about this. It is substantial.

SIEVER: Would a sulfur isotope fractionation from the sulfur contained in that be expected?

MILLER: I think the chances of getting a sample of cysteine or methionine is particularly bad.

WOLMAN: I don't know how stable cysteine will be.

MILLER: Yes, it will decompose. It is particularly bad because of decomposition.

BADA: Are there more comments?

Kerogen

BARGHOORN: There's one thing about this contamination. There is a preference for L-amino when ancient rocks from the Fig Tree are used. Are these free amino acids that come out? What comes out of this junk that we call kerogen, when it is cooked up? Also the L-amino acids? Most of the amino acid is bound in this kerogen garbage.

BADA: In an acid hydrolysis it is cooked up.

BARGHOORN: But are the L-amino acids coming out of the kerogen?

BADA: I don't know.

BARGHOORN: This is why they would be bound. This is very important, because this stuff has to be bound within 30,000 years.

BADA: Yes. Does anybody here know of anybody who knows?

ABELSON: It is a good question.

BADA: Most people just look at the extractable amino acids from this chert.

SIEVER: How do you extract from the kerogen?

BADA: It would have to be hydrolyzed in acid.

ABELSON: Kerogen is interesting. Add sixth normal hydrochloric acid to kerogen and cook it for a day; a certain amount of amino acid comes out. Put it back, cook it for another day, and about ¾ amino acids comes out the second day. After about five weeks of this procedure cook it some more, and you still get some amino acids out of it. There is some kind of peculiar binding, God knows what, of amino acids in the kerogen association.

MILLER: Has base hydrolysis ever been used on this?

ABELSON: I haven't used it.

SCHOPF: John Hunt and Egon Degens (1964; Degens, 1965, p. 254-255) did a comparable experiment five or six years ago. They alternated base-acid hydrolysis of kerogens and some humic acid materials. They reported very similar results; you could keep getting a yield of amino acids, suggesting that they were bound in some way.

BARGHOORN: This was in a recent core, not a deep sea core, wasn't it?

SCHOPF: The humic acids were Recent, from the San Diego Trough; kerogen was from the Woodford Shale, of Mississippian age.

ABELSON: Yes, this wouldn't be ancient material.

LEMMON: But has no one looked at the optical rotatory power of these hydrolysates?

ABELSON: Not to my knowledge.

LEMMON: It is easy to do.

BADA: Do you mean the study of hydrolysate in a polarimeter?

LEMMON: Yes.

BADA: There would not be enough material.

ABELSON: But it would be easy (we may even have done this) to simply look at the ratio of allo to isoleucine.

LEMMON: Sure.

USHER: In your rate of racemization experiments have you studied different concentrations?

BADA: Yes.

USHER: What is the effect of solvent polarity? Could mixed solvents be protecting it against racemization?

BADA: Would you expect the racemization rate to be faster in ethanol?

USHER: You must get it in solution.

BADA: I don't know, it is possible. That is strictly an exercise in physical organic chemistry.

USHER: Not really. The rate in the absence of water may be faster not because of catalysis, but because of protection in active solution by the hydrogen bound carboxylate with the amino group, which stops them reacting.

BADA: Okay. That may be an explanation of what we have seen. I think I can neglect the water reaction, that the hydroxide ion is pulling off that proton in that water. We can discuss this later, but there are a couple of water reactions which aren't required in the kinetic expression. In fact the rates of these required reactions should be fastest by several orders of magnitude, based on the fact that the water reaction probably can be neglected.

PONNAMPERUMA: On the hydrolysis of the kerogen — is this demineralized, Dr. Abelson?

ABELSON: Yes.

PONNAMPERUMA: So removing the minerals with HF, is there still trouble getting out amino acids?

ABELSON: Yes.

ORGEL: Does it all dissolve in the end?

ABELSON: I don't know, you get tired after a while! There must be better ways to live life than to keep hydrolyzing!

BARGHOORN: I think Jeff Bada has more to say.

BADA: Very briefly. I have come across something quite unexpected that maybe should be seriously considered in the future. I was looking at phenylalanine in the presence of formaldehyde. In one experiment I buffered the phenylalanine to pH 7.6, and in a hundredfold excess of formaldehyde. The amino acid concentration was around 10^{-4} molar, and formaldehyde was around 10^{-2} molar. I cooked this up at 118°C for three or four days, because the rate of decomposition of phenylalanine is very slow. At 118° for three or four days the per cent decomposition of phenylalanine (in the absence of formaldehyde) is about a 0.01 per cent. However, in the presence of formaldehyde, I found that phenylalanine was decomposed very rapidly.

I then did further experiments varying the formaldehyde concentration. It appears that as long as formaldehyde is in excess, phenylalanine is very rapidly decomposed. Although some sort of Schiff's base or hydroxy-methyl substituted amino acid, which very rapidly decarboxylates can be postulated, we are not clear what the decomposition mechanism is at the present time.

Apparently formaldehyde catalyzes the decomposition of at least some amino acids, which has implications for the use of formaldehyde as an important prebiological molecule in the synthesis of more complicated monomers. The concentration of formaldehyde in the primitive ocean must be considered in order to not completely decompose amino acids. From preliminary data it seems that as long as the concentration of formaldehyde is a factor of ten less than of the phenylalanine, the decomposition rate is very slow. But at comparable or excess concentrations of formaldehyde, phenylalanine rapidly decomposes.

MILLER: What time scale do you mean when you say "very fast"?

BADA: I don't know the activation energy but at elevated temperatures ($\sim 118°C$) it is on the order of an hour or less.

PONNAMPERUMA: I think it is unreasonable to expect formaldehyde to accumulate under any circumstances, since it would polymerize.

MILLER: Would a steady state level of 10^{-4} molar formaldehyde be unreasonable?

PONNAMPERUMA: At basic pH?

MILLER: Yes, pH 8. A 1.0 molar solution will give sugars, 0.01 molar – may give sugars too, but what about 10^{-4}?

PONNAMPERUMA: What concentrations did you find in your early experiments?

MILLER: It depends on the experiment, but between 10^{-3} and 10^{-2} M, I believe.

PONNAMPERUMA: In the electric discharge experiments?

MILLER: In silent discharge experiments, large quantities of formaldehyde were formed, but with the spark discharge there was relatively little, because the formaldehyde is bound up with glycine nitrile and glycolic nitrile.

SAGAN: Are a few parts per million of formaldehyde enough to cause decomposition of phenylalanine?

BADA: What is the solubility of formaldehyde?

MILLER: It is a bit more than volatile than water. At 25° the vapor pressure of water is 23 mm; a 10^{-4} M formaldehyde solution would have 2 ppm of formaldehyde in the gas phase above it.

SAGAN: Because parts per million is still enough to be opaque in the long wavelength UV region that I showed before. It is possible, if I understand Stan, to have sufficiently little formaldehyde in the ocean to avoid Bada's phenylalanine problem, and still have enough formaldehyde in the atmosphere to be a long wavelength photon acceptor.

MILLER: But more formaldehyde is needed for sugar synthesis if they are to be made via Buterow polymerization (Reid and Orgel, 1967).

What is the limiting concentration for sugar synthesis?

HULETT: About 10^{-4}, as I recall.

ORGEL: Thermodynamically. I think experimentally it is 10^{-2} M.

PONNAMPERUMA: 10^{-3} M.

MILLER: Is that in the clay mineral?

USHER: What happens to the phenyl group?

BADA: Do you know a normal method of titrating ammonia, using the Conway diffusion cell? The phenyl constituent diffuses here just as ammonia does, which suggests it becomes phenethylamine. But since I have never put it on a gas chromatograph I don't know whether it is phenethylamine.

USHER: So we invoke the Belmont Manifesto, and you tell us later what it really was!

SANCHEZ: Would phenethylamine show up on the amino acid analyzer?

MILLER: If a relatively high concentration of formaldehyde is needed, namely 10^{-2} or 10^{-3} M, that appears to be enough to wreck the amino acids. As Jeff says, the data is preliminary, and all the products aren't known, but it is of concern. Amino acids are made with difficulty then they all get decomposed by formaldehyde.

SAGAN: Yet a lot of formaldehyde is needed to make sugars.

ORGEL: Don't we have a much worse problem in that the amino acids and sugars aren't stable together anyhow? I don't think the formaldehyde situation makes the problem any worse.

MILLER: Yes, that's another point Phil Abelson has brought to our attention several times. The amino acids and sugars interact and make all sorts of colored products, because sugar is an aldehyde. The Browning or Yellier's reaction may also do a good deal of damage to amino acids in the primitive ocean.

ABELSON: About 95 or more per cent of the reduced carbon in sediments is in the complex form of kerogen. This is a name for organic matter that does not dissolve in any reasonable solvent, but can only be dissolved by oxidation. It is like some very high molecular weight polymer. Nevertheless, we must be interested in it, because it is the resting place of various carbon skeletons, and in it resides the possibility of looking at compounds that come from very old life. Kerogen is very difficult to study. How can you find out anything about a black, insoluble goo?

Tom Hoering (Carnegie Institution of Washington, Geophysical Laboratory) has tried to get at it with a mixture of reduction and partial oxidation. Ultimately he gets pieces that can be identified via GLC mass spectrometry. Kerogen turns out to be incredibly complex in terms of the various carbon skeletons that come out.

We met limited success in approaching the study of kerogen through the analysis of kerogen and its probable precursor, humic acid. One of our best approaches was to study these materials by making them artificially.

Living matter contains lipids, carbohydrates, and proteins. Lipids aren't all that reactive, but what about carbohydrates and proteins?

It has been long known that a simple mixture of glucose and amino acids give a Maillard reaction and complex products. We have prepared many humic acids and, indeed, kerogens from glucose plus amino acids.

If equimolar mixtures of the two are heated (there is nothing sacred about equimolar — you can use various proportions of glucose and amino acids) at the end of the day there is some precipitable humic acid. If the reaction is allowed to continue for some weeks, you can observe variable results.

For example, after a week or so of heating glucose and lysine apparently bridges are formed and behaves just like kerogen.

Although the reaction of glucose and amino acids constitutes a model, one can not be sure that the model is a good one. We are never completely certain; but we proceed on the basis that if it has four legs, wags its tail, barks, and chases automobiles, it resembles a dog — perhaps it is another canine species; if we pile up various concordant criteria, we are more and more convinced that it is dog-like. We start with reasonable chemicals, cook them together and get substances that have CN ratios that are the same as natural materials.

It is also interesting that these artificial "kerogens" have a similar hydrolysis behavior to that I mentioned earlier, they slowly release amino acids. We also isolated some of the artificial kerogen, and studied the reactivity of free amino acids with it. The artificial material behaves in very much the same way as natural kerogen. Comparing their electron spin resonance we found the artificial and authentic materials are similar. We also found that the reactivity is sensitive to exposure to oxygen and to strong reducing agents in both cases.

In all tests we have tried so far the artificial humic acids and kerogens are remarkably similar to material isolated from sediments.

LEMMON: Dr. Abelson, is this always a mixture of glucose and all the amino acids or just certain ones?

ABELSON: We have formed humic acid-like materials from virtually all of the amino acids and mixtures of them to glucose.

LEMMON: Which comes out most like kerogen in all respects?

ABELSON: They are all like kerogen to a certain degree, but, for example, glycine or alanine, over six months' time yield very little kerogen. I believe this reflects an hydrophilic combination. It is much easier to get an artificial kerogen from more hydrophobic amino acids, for example, phenylalanine and especially lysine.

There also seems to be some kind of additive effect, in that a mixture of amino acids apparently makes kerogen faster than the individual ones. In addition to the straight reaction of the glucose and amino acids, there seems to be some interaction among the R groups.

FERRIS: Can you tell us more about the reaction conditions for this?

ABELSON: It seems impossible to miss on this one!

We usually run the reactions at pH 7 or 8 and 100°C but this reaction will occur at pH 4, pH 9, and also (though very slowly) in the refrigerator.

MARGULIS: What are the lipids involved?

ABELSON: None. This is just amino acids.

ORGEL: Does it make any difference if air is excluded?

ABELSON: Air does have a role, but the reaction rate is about the same under nitrogen.

SAGAN: Does the amount of water present make a difference?

ABELSON: We have chosen to do these experiments in $1.0 M$ or $0.10 M$ solutions but the reactions go on in more dilute solutions.

ORGEL: If this were evaporated down you would probably get solid material too?

ABELSON: We haven't done that.

Table 7 shows a series of runs in which we exposed three amino acids to some kerogen. At 25° there is essentially a complete disappearance, for example, of arginine and disappearance of half cystine and phenylalanine. The residual levels of the dibasic amino acids are way down. That is true throughout the temperature range from room temperature to 100°C.

TABLE 7. REACTION OF PEPTIDES WITH KEROGEN*

	Start	1½ hr	6½ hr	1 day	4 days
Glycylleucine + kerogen at 110°C, pH 8.8					
Glycine	0	1	3	5	7
Leucine	0	1	5	8	8
Glycylleucine	100	84	49	18	6
Leucylglycine	0	...	tr	1	4
Leucylglycine + kerogen at 110°C, pH 8.8					
Glycine	0	2	4	7	11
Leucine	0	tr	1	2	3
Leucylglycine	100	86	57	40	15
Glycylleucine	0	1	1	2	4

*Data expressed as percentage of original peptide. 6 N HCl hydrolysis of the 4-day kerogen residue yielded less than 10% of the glycine and leucine originally present. Tr, trace.

My point here is that important abiotic mechanisms exist for destroying amino acids at 25°. Extrapolation from the times and temperatures indicate that it would only be some years before these amino acids disappeared even at 0°C.

ORGEL: What about aspartic acid?

ABELSON: Aspartic acid is the least reactive, and so we have used aspartic acid as our internal standard.

FERRIS: Is it just because it doesn't stick because of an acid-base property, do you think?

ABELSON: Notice that the hydrophobic phenylalanine is more reactive, so in part there is a hydrophobic-hydrophilic distinction. In addition, the dibasic amino acids, lysine and arginine, are reactive. We are pretty sure that the reaction is via the amine group. One way to show this was to react arginine with kerogen and observe the change in the carbon-nitrogen ratio.

BADA: Is it decarboxylated?

ABELSON: In this particular instance we really don't know what we were doing. We were just following the carbon-nitrogen ratios.

SANCHEZ: Where does this particular sample of kerogen come from?

ABELSON: This happens to have been isolated from a recent marine sediment off southern California, but we have done this with about ten different kerogens from all over the world. They aren't identical in their reactivity, the picture shown here is generally true.

SANCHEZ: Does synthetic kerogen do the same?

ABELSON: Yes.

ORGEL: What happens with the naturally occurring mixture of amino acids in roughly the right ratios, heated up with glucose and hydrolyzed? Does it give a picture similar to that of the natural kerogen?

ABELSON: We have not done precisely that.

PONNAMPERUMA: What method of extraction do you use for recovery?

ABELSON: Several. One is to make the pH 2, and analyze the supernate. We also take the residue and treat it with the 6N HCl. to see if we can get anything more. We get minute amounts of additional amino acid from the 6N HCl treatment.

BADA: This stuff has been preleached with HCl?

ABELSON: Certainly. We don't want to have a background of free amino acids coming along with the kerogen.

ORGEL: Which three amino acids come out of natural kerogen?

ABELSON: All, including, incidentally, cysteine.

ORGEL: So this behaves differently.

ABELSON: Remember, this experiment was the incubation of kerogen with amino acids. Now, you asked what happens on long-time cooking with 6N HCl. The reaction of free amino acids with kerogen is damped down. This is the reaction now occurring at the pH of sediments, though it goes almost as fast at pH 4. It isn't highly pH sensitive. But once the solution is made 6N HCl, the reaction is slow.

SIEVER: Phil, a good place to look for this would be a peat bog or swamp, where a lot of your conditions are met (granted usually the pH is around 5 or a bit below). All the humic acids should be there (of course bacteria are there, and that is a problem) but your process should be eating up all the free amino acids.

ABELSON: We have had kerogen, humic acid — from the Black Sea, the Persian Gulf, Venezuela and southern California. We have also had it from some loam soil. Everything we have isolated and exposed to these amino acids reacts with the amino acids.

ORGEL: Do bacteria gobble the artificial kerogen up in turn?

ABELSON: No, apparently not very readily, because it is preserved for millions of years.

ORGEL: But in places where it is preserved are there lots of bacteria for a million years?

ABELSON: I mentioned briefly our observation of similar material — mud from the depths of southern California. This was pulled up aboard ship, frozen, and two years later we wanted to see what was going on. We suspended this mud, gave it a bit of glucose, and within a day all the glucose had been consumed! This particular mud apparently has a very voracious group of microbes, and the fact that the kerogen too hadn't disappeared implies a certain resistance to bacterial attack.

BADA: Was this mud from the Santa Barbara basin?

ABELSON: Yes.

BADA: This probably has one of the highest sedimentation rates (on the order of 2 or 3 cm a year) found anywhere along the California coast.

ABELSON: We have had it also from the San Pedro basin and it makes no difference.

BADA: That's where Los Angeles dumps all its sewage!

SANCHEZ: Presumably you might be able to do the opposite: feed sugars to kerogen and see them being absorbed.

ABELSON: I don't know, it might happen.

SCHOPF: Have you ever found any naturally occurring kerogens that were saturated and couldn't absorb any more amino acids?

ABELSON: No.

MARGULIS: When you hydrolyze either the natural or the synthetic kerogens, do the amino acids come off in some sequence, or at random?

ABELSON: No, a little of all of them are present.

BARGHOORN: Would you repeat the exact history of what is shown in Table 7 (p. 149)

ABELSON: This is kerogen isolated from mud, followed by the usual demineralization. At the time we ran these experiments this material was 98 per cent organic matter.

Table 7 shows a combination experiment in which we exposed some peptides to kerogen. These results don't exactly show it, but we found if a glycylleucine is added and exposed to kerogen, subsequently hydrolyzed, we get leucine off predominantly. In other words, the amine group was bound irreversibly, and not the peptide linkage. When the combination was hydrolyzed, off came leucine.

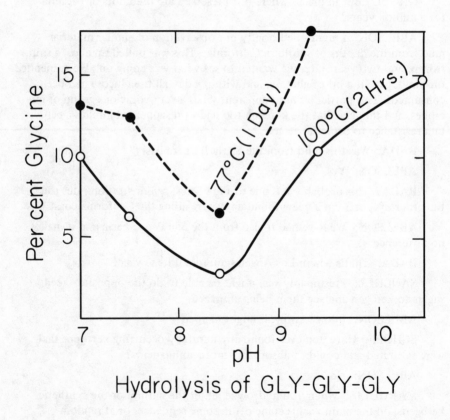

Figure 11. Percent glycine yielded by hydrolysis of triglycine polymers as a function of alkalinity at 77° and 100° Centigrade

In the course of these experiments we found that incubation of peptide after a long time yields a new entity, as shown in Table 7. That is, in the upper group, we start out with glycylleucine, after four days we have some leucylglycine. At the bottom we start out with glycylleucine, and end up with some leucylglycine. This is not so very strange, because we must be making a diketopiperazine − in part − and then when that diketopiperazine breaks up, it doesn't remember what its initial parents were.

TABLE 8. PERCENTAGE RECOVERY OF AMINO ACIDS AS A FUNCTION OF TIME AND TEMPERATURE FROM KEROGEN-AMINO ACID MIXTURES*

Amino Acid	25°C				52°C		80°C				110°C			
	1 hr	97 days	8 days	83 days	1 day	2 days	8 days	6 hr	2 days	4 days	8 days
Lysine	95	13	0	0	6	0	0	0	0	0	0
Histidine	90	10	tr	tr	11	0	0	5	0	0	0
Arginine	90	4	0	0	2	0	0	0	0	0	0
Threonine	98	64	77	27	65	65	36	60	24	13	6
Serine	98	71	79	37	68	69	42	66	31	18	9
Glutamic acid	98	97	100	95	87	90	90	91	76	61	60
Glycine	99	43	64	38	77	57	35	59	32	24	18
Alanine	99	80	81	61	61	60	52	65	35	22	15
Half cystine	29	0	0	0	0	0	0	0	0	0	0
Valine	83	80	62	57	61	54	26	50	19	9	3
Methionine	61	43	31	6	35	25	14	10	0	0	0
Isoleucine	82	66	47	38	46	36	11	33	11	0	0
Leucine	88	60	32	23	40	21	2	16	0	0	0
Tyrosine	36	19	9	3	29	6	0	10	0	0	0
Phenylalanine	30	12	2	1	16	2	0	7	0	0	0

*0.2 μM of each amino acid originally. Data expressed as percentage recovery with aspartic acid normalized to 100%.

MILLER: Was a control of the leucylglycine run in aqueous solution?

ABELSON: We did those separately, and we found that we'd get essentially the same results. So forget about the presence of the kerogen. That just happened to be the way we stumbled into this thing. The significant matter is that if you start out with dipeptide, there is a preference. Leucylglycine is preferred to glycylleucine.

I don't know whether this is consistently true for amino acid sequence, but if this dipeptide is permitted to equilibrate, you end up with more leucylglycine than glycylleucine.

WOLMAN: It was shown that the hydrolysis of the diketopiperazine of prolylglycine depends on the pH at which it is done. Mild acidic pH yields prolylglycine, and mild basic pH gives you glycylproline.

MILLER: Is that Brenner's work?

WOLMAN: No, This is Russian work, from twelve years ago (Poroshin *et al.,* 1958).

ABELSON: This whole business of pH effect and also this intrinsic tendency for one amino acid to succeed another must be studied further.

SANCHEZ: Dr. Abelson, I'm not sure I understand Table 7. (p. 149)

The footnote implies hydrolysis of the four-day sample gives a relatively low recovery of the original glycine and leucine. I'm surprised the recovery isn't much higher, in view of the rates of release of your reaction conditions.

ABELSON: Since kerogen tends to react irreversibly with materials, there is a disappearance of everything. Note some leucine and glycine is freed. At the end of four days, there is no accountability, because most material has been taken up in the kerogen. The moment amino acids or whatever are released, the kerogen reacts with them.

SANCHEZ: For example, it shows that 85 per cent of leucylglycine is absorbed. Would you not expect the major proportion of the glycine to be released on acid hydrolysis?

ABELSON: These are our results.

SCHOPF: Do Precambrian kerogens characteristically absorb free amino acids?

ABELSON: I don't know whether Gunflint chert has Precambrian kerogen in it or not, but our work on the reactivity of kerogen started when we looked for free amino acids in the Gunflint chert. We added some norleucine to have a nonbiological tracer and in processing took it down to dryness. Several things happened, but eventually we discovered that norleucine had essentially disappeared.

Because we found kerogen in the Gunflint chert was reactive with amino acids we pursued this work.

SCHOPF: I may not be visualizing this correctly, but this seems to me surprising. Presumably there is some finite number of sites in kerogen to which amino acids could become attached in some way. If groundwaters move through a sediment over geologic periods of time, at some point the kerogen should become saturated.

ORGEL: But it may continue to pick up sugars as well, which might allow it to keep accumulating.

ABELSON: Don't underrate the possible capability of the kerogen to digest and alter material it reacts with.

BARGHOORN: Did you demineralize the Gunflint chert with HF?

ABELSON: Yes.

SOFFEN: Does kerogen show a preference for one or another of the isomers? Does it distinguish L from D amino acids?

ABELSON: I don't know.

BADA: Did you use a 24-hour hydrolysis?

ABELSON: More or less, overnight hydrolysis.

BADA: Did you use 6N boiling HCl?

ABELSON: It's sealed under nitrogen at 108°.

BADA: Why didn't the leucylglycine hydrolyze?

ABELSON: Kerogen was treated with 0.01N HCl, and the supernate put on an amino acid analyzer.

MILLER: Implying that the peptide analyzed in the analyzer was not hydrolyzed.

ABELSON: Yes.

HULETT: And then does the later 6N HCl hydrolysis step give the extra ten per cent listed there at the bottom?

ABELSON: Yes. The data expresses percentage of original peptide. We then take 6N hydrochloric acid hydrolysis of the 4-day kerogen residue to show there was no adventitious adsorption. All we have show here is that kerogen was a good absorber.

PONNAMPERUMA: What is the reaction process?

ABELSON: Kerogen is extremely reactive. Certainly one important reaction route is via the amine group; but that is not necessarily the only way in which kerogen reacts.

ORGEL: Have you done any straightforward chemistry to look for aldehyde groups? It sounds like the chemistry of aldehydes and amines.

SANCHEZ: Perhaps kerogen could be titrated with sodium bisulfite or something. It should have the same equivalents for amino acids.

ORGEL: The aldehyde reactions may involve the formation of a Schiff's base followed by oxidation-reduction, to give in this case, methylamine, and the aldehyde of the amino acid.

BADA: Or an alpha ketoacid.

ORGEL: Yes. There seem to be one or two pyridoxal type transaminase, which formaldehyde is known to do.

PONNAMPERUMA: Dr. Abelson, in your work you identified some fatty acids, some from the acid treatment of kerogen. Is there a possibility that these may play a role? The mechanism may not be purely chemical in nature.

ABELSON: I am not convinced that the hiding of substances in kerogen was entirely chemical. Because of its complexity, it might very well have adsorptive capabilities. For example, some lipids may still be there somehow adsorbed pretty tightly to kerogen.

ORGEL: Are there examples of a substance so difficult to extract being adsorbed that it does not form covalent bonds?

ABELSON: I don't know.

WOLMAN: Did you every get a complete adsorption of amino acids? Did you ever saturate the kerogen such that no more amino acids were adsorbed?

ABELSON: We have never really looked.

If amino acids exposed to kerogen are not completely taken up, why weren't they? Perhaps sufficient time had not elapsed for the kinetics, or the sites are not being available?

Although we have performed uptake experiments at different concentrations, we have made no effort to saturate with excess amino acid. At lower concentrations the percentage disappearance of the amino acids is greater. There may be a spectrum of reactive sites.

LEMMON: Unless glucose keeps building kerogen there must be some saturation point.

ABELSON: I bet if I incubated arginine long enough, all the arginine would disappear. Somehow I feel in the process of reacting more reactive sites would appear.

LEMMON: Are you suggesting a mechanism to polymerize arginine?

ABELSON: No! As we found out in our first travail in organic chemistry, a property of nature is to make black tar! This has been an investigation of a special form of black tar.

I want briefly to discuss our triglycine results (Fig. 11). We were interested in how fast it would fall apart. At 100° it decomposes fairly fast, and even at 77° there is a fair amount of change. If you start with good, clean triglycine, and look for glycine itself — you are starting from zero background. Glycine is slowly built up until it can be detected quite readily. It was amusing to us that the optimum stability of this peptide was in the region pH 8 to 9.

MILLER: What buffers did you use?

ABELSON: Phosphate buffer, among others. We checked that with the pH meter at high temperatures.

HULETT: Is that the per cent glycine (Fig. 11, p. 152).

ABELSON: No, per cent glycine free. If you split a triglycine peptide bond you get glycine.

MILLER: Was diglycine detected?

ABELSON: Yes.

USHER: The test would be to use tetraglycine, to see if you get diglycine before you get monoglycine.

ABELSON: This particular experiment is easy to do. The advantage in the diketopiperzine is that it doesn't bind up on Dowex, so you can get it through a resin bed.

ORGEL: There is a peptidase enzyme that takes amino acids off two at a time.

USHER: I didn't know that.

SCHOPF: This sort of kerogen formation can be taking place in recent sediments, and in peat bogs, but is there any reason that it should not occur within rocks?

ABELSON: The constituents must get together.

ORÓ: From three slides you presented what are your conclusions in relation to the problem of this conference?

ABELSON: I am struck by how tough the problem is of the origin of life. Part of the reason for these investigations is to place some boundary conditions on it. An understanding of the limitations may actually force us into some particular kind of mechanism.

The moment Stan Miller did his work the tendency was to say: we have made amino acids, therefore life is inevitable. In a certain way this conclusion destroys the origin of life problem: if it's inevitable, why work on it? However, if it is really difficult to avoid these various pitfalls of disappearances, degradations, instabilities – then we may get guidance in attacking the problem. For example, I feel that a consequence of the tendency of amino acids to disappear is that we must think in terms of harnessing sunlight in some way that leads to copious production of amino acids.

This forces a way of thinking that may turn out to be fruitful; apparently in order to fit these boundary conditions, life may have originated in a highly specified way.

MILLER: Another way is to say this is that it is almost as important to investigate the stability of a prebiotic compound (e.g. cysteine) as it is to synthesize it under prebiotic conditions.

ORGEL: We need to think about the compatibility of the amino acids, also.

PONNAMPERUMA: Should some effort be made to search for pre-biotic kerogen on the earth? Presumably most kerogen we work with has resulted from organic matter accumulation through biological processes.

MARGULIS: Yesterday we heard that these rocks are porous. Couldn't kerogen be produced steadily over the past billions of years?

BARGHOORN: You use the term "kerogen" which is the end product referring to something highly reactive. Could you expand on what you mean by highly reactive, in other ways than in reaction with amino acids?

ABELSON: I once had some kerogen which I had painstakingly isolated from Green River shale. I was so incautious as to warm it up to 75°C in the presence of air; whereupon it all burned up. So it's reactive in that sense!

BARGHOORN: It is also reactive in that if Gunflint kerogen is washed and warmed up thoroughly, and let set, it will support fungi and bacteria.

ABELSON: This is something else again. These aerobic creatures do marvelous things with the utilization of organic carbon. This observation is a testimony more to the biological capability of fungi than to the reactivity of the kerogen.

SIEVER: Have any of your prebiotic synthesizers backreacted the for-maldehyde and amino acids with the red-brown material, turned off the spark, or the UV, and let it sit to backreact? If so, is there any similarity at all between the red-brown and the kerogen?

The instability could be seen: the amino acids and formaldehyde would disappear back into the red-brown stuff. Has anybody done that?

MILLER: If it sits in the refrigerator for a long time, the results are about the same.

ORO: I am still trying to get to the positive aspects of Dr. Abelson's slides. In the first (Table 7, p. 149) I saw the amino acids which are most stable, and are found relatively less destroyed towards the end of the reaction. I believe aspartic, glutamic, glycine and alanine are some of the ones that remained relatively unchanged. Interestingly enough these are the same amino acids which are easier to form under a number of experimental prebiotic conditions. There is a ray of hope that perhaps the ones we make easily are also more stable; and therefore could be around for a longer time on the primitive earth.

The second slide (Table 2, p. 153) gave information on the rate of absorption (or combination of these amino acids into the "kerogen" polymer. I also look at it positively, assuming there is a possibility of release of the amino acids once they are incorporated. Some of these amino acids may be retained in the polymer for availability at a later time.

A possible mechanism of interaction may involve transamination reactions between keto acids and amino acids. This reaction may also be related to earlier experiments of amino acid synthesis with formaldehyde (Oró *et al.*, 1959). We have demonstrated a simple non-enzymatic transamination between amino acids and keto acids by means of some metallic ions (Doctor and Oró, 1967, 1969).

MILLER: What was your solvent system?

ORÓ: Water. This transamination is analogous to that in which amino acids react with pyridoxal forming Schiff bases. Our system contained no pyridoxal. We tested a number of metal ions and if I recall properly, copper, iron, and aluminum were some of the best.

It is the same general prebiotic chemistry problem but may imply kerogen activity as a proto-catalyst. It may facilitate transamination reactions in which the amino group of an amino acid interacts with a carbonyl group on kerogen to form a Schiff's base. Now the last step is that amino acids may be released by a process of transamination, by cleavage of that C=N bond. If, however, kerogens are irreversible to further exchange processes, the amino acids may not be available any more.

This answers the question of whether these polymers can grow indefinitely. In principle they can, because you start with a carbonyl group, put on an amino group; you end up with a carboxyl and another amino group in which, if a transamination reaction occurs, the amino group may become a keto group and pick another amino acid, and so on. It is not an impossibility that this could have been an "infinitely" growing polymer.

BADA: Does the alpha keto acid persist very long? I thought they immediately decarboxylate to aldehydes.

SCHOPF: In some ancient Precambrian sediments several billion years old, kerogen can not be growing indefinitely. If the organic microfossils are composed of kerogen, this can be seen at an ultrastructural level, as well as a light optical level. Apparently the microfossils are not getting bigger by accreting organic matter.

ORÓ: Who in their right mind thinks a rock can grow?

USHER: I'm not sure. An alpha amino acid in a peptide linkage by its carboxyl end, can be converted – the amino group to a keto group – then another amino acid can be brought in, but now there is a 3-carboxyl group that can't grow any more.

ORÓ: He stated lysine was one of the best amino acids but no matter which amino acid you have, the epsilon amino group of lysine will be attached.

USHER: No, lysine is a special case. Such kerogen should be lysine rich after ten years.

ORÓ: Apparently lysine is one of the best.

LEMMON: Arginine was good too.

ORÓ: Arginine is dibasic too. I'm not predicting anything but just offering a possibility for observations that are not readily explicable.

If I understood correctly your third slide (Fig. 11) showed the hydrolysis of triglycine at different pH's. It seems that around pH 7 and 8 is the minimum.

ABELSON: Yes, at pH 8 the rate of hydrolysis is minimum.

ORÓ: That is another very interesting observation. This means that the particular polymer, triglycine, is quite stable at pH's between 7 and 8 which are reasonable for the terrestrial lithosphere or hydrosphere. The stability of the glycine peptides is compatible with our model that has the early ocean about pH 8.

I was just trying to be God's advocate in this!

SAGAN: Well, referring to Jeff Bada's remarks: if we have values for half-life of amino acids against destruction, and decent production rates, we are in a position to calculate the steady state concentration of amino acids in the primitive oceans. Yesterday I presented a production rate from our long wavelength ultraviolet amino acid synthetic experiments: 200 kg per square centimeter column for 10^9 years, assuming no destruction. Jeff's racemization half-life is about a million years at $0°C$. He stressed that the thermal degradation half-lives were longer than the racemization half-lives. Let us take 10^6 years as a typical half-life for amino acids, resulting in the reduction by a factor of a thousand the 200 kg per cm^2 column. This quantity mixed in an ocean of water of 2×10^5 gm per square centimeter implies a 0.01% solution of amino acids by mass. The degradation products of the amino acids are also organic compounds of some sort, and since the production rates talked about apply only to the amino acids, steady state concentration of organic compounds might even be higher, perhaps several tenths of a per cent.

BADA: You probably forgot the natural mechanism of removal by sediment. In the present ocean, the major mechanism of removal is by bacterial oxidation or other biological utilization.

SAGAN: It does not apply to the primitive ocean.

BADA: Yes, but nobody really knows now what proportion of the organic material is removed by sediments. It is probably 100 or a 1000 to 1.

MILLER: In favor of the sediments, or of the ocean?

BADA: In favor of the organisms. Particulate organic matter is different. Most dissolved organics are removed by organisms and the rest is eventually removed by the sediments.

MILLER: Don't clay minerals absorb organics fairly well?

BADA: Yes. Montmorillonite and illite pick up amino acids quite easily. In a primitive ocean the decomposition along –

SAGAN: The question is what is the half-life of amino acids in an abiotic ocean against loss by sedimentation and other processes?

HULETT: At least some photodestruction in the atmosphere and in the ocean must have occurred until there was enough tar to make the ocean completely opaque.

SAGAN: Photodestruction in the atmosphere is included in our estimate of the production rate. I used our laboratory value.

HULETT: Yes, but production by shock wave –

SAGAN: No! This has nothing to do with the shock wave results; this estimate was based on our ultraviolet experiments.

HULETT: But still have you the ultraviolet wavelengths that are destroying them at the same time in your apparatus? Have you really simulated the solar spectrum?

SAGAN: Why is the entire solar spectrum important?

HULETT: Edwards found if he kept his synthesis going, the actual production of organic material got less.

SAGAN: The question of what spectrum was used should not be confused with the problem of the continued irradiation of the products. Those are different, right? Our answer is that the circulation in our system is slow enough that the molecules that are synthesized before solution in the aqueous phase certainly got a very large ultraviolet dose.

BARGHOORN: What is the ratio of organic particulates to organic dissolved materials in the present ocean?

BADA: Ten to one dissolved.

But I might clarify this 3400 year age for dissolved organic materials in the deep ocean. It appears that most of the dissolved organic carbon is amorphous humic acid, whatever that is. The dissolved materials certainly are not amino acids, carbohydrates or other common biochemicals. This has suggested that most compounds like amino acids and sugars are removed rapidly, and this remaining stuff is mixed up. In fact the 3400 years reflects the rate of removal by the sediments of this amorphous material.

SAGAN: But you have no idea by how many orders of magnitude this 3400 year estimate figure would be expected to rise if there were no life in the ocean.

BADA: No. It may go up by a factor of ten; I can't imagine it going up by much more, but I am simply speculating. Dissolved organic, amorphous material in the deep ocean, can't be used by organisms anyway.

If the residence time were longer – say a million or ten million years – it should be accumulating in the ocean like crazy.

SAGAN: The question is the ratio of the rate of destruction biogenically to the rate abiogenically. By "destruction" abiogenically I mean removal from the ocean.

BADA: The major mechanism of removal of the dissolved organic fraction in the deep ocean right now I would estimate is probably through the sedimentary column.

Approaching the surface of the ocean, the spectrum of dissolved organic material probably changes, there should be more amino acids and carbohydrates which are being removed rapidly.

SAGAN: Right. And for amino acids and carbohydrates, what is the ratio of abiogenic to biogenic removal?

BADA: We don't know.

SAGAN: It is possible that you will end up with a half-life against removal by abiogenic processes of millions of years.

BARGHOORN: How much organic matter is the oceans could be of terrigenous origin and of phenolic nature?

BADA: Probably a lot.

SAGAN: Does "terrigenous" mean from the surface?

ORÓ: Land versus marine.

BADA: Also, a considerable fraction probably comes from lignin-type polymers (from woody land plants) and algae.

MILLER: Are you trying to calculate the 1.0% or 0.1% concentration in the ocean?

SAGAN: I'm not trying to get a particular answer.

MILLER: If you dissolve in the present ocean the whole sedimentary carbon column you get 1.0%.

SAGAN: 2.0%.

MILLER: Wouldn't it be just as effective to pick the fraction of the total 3000 gm/cm^2 that was organic compounds in the primitive ocean?

SAGAN: Maybe the question for the geologists is: Is it possible that much carbon in the first billion years of earth history was unavailable for organic syntheses, say in the form of carbonates?

SIEVER: I would estimate not any more than now. To produce carbonates reduced carbon is just being oxidized. Presumably in the past you were producing reduced carbon as well as CO_2, and there really is no mechanism for oxidizing it. Well, there might be some mechanisms from oxidized metals, I thought we could discuss this later in the afternoon.

Precambrian Carbon

MILLER: Is there any possible way of avoiding calcium carbonate precipitate in the primitive ocean? We talk about a reducing atmosphere but, for example, if urea is made it would have been hydrolyzed to CO_2.

SIEVER: As long as there is CO_2, there will be carbonate. The only question here is the ratio of CO_2 to reduced carbon in gases.

SAGAN: Won't that be the same as the ratio of carbonates to organic carbon in the ocean?

SIEVER: This depends on the amount of hydrogen in the atmosphere, which controls the ratio between CO_2, CO and CH_4.

BARGHOORN: Is not the percentage of carbonates in older Precambrian rocks conspicuously low?

MILLER: Is that explained by greater weathering, or that there was not as much carbonate produced?

SIEVER: The assumption has been that the early Precambrian rocks were weathered more, but it is open to question.

ORÓ: Ray, tell us please about the source or origin of the carbonatites. Are they primary or not?

SIEVER: I don't know a damn thing about carbonatites.

ORÓ: Some authors have supposed them to be of primary origin. Although they may not be, it would tend to answer your question on the quantity of carbonate and inorganic carbon.

SAGAN: How?

ORÓ: If a geologist knows all the places on earth where the carbonatite formations are — this would give a figure of the amount of carbon. I don't know of any amount of reduced carbon, since it is supposedly prebiotic — it must have all been oxidized. To calculate this good evidence is needed that the carbonatites are primary.

SIEVER: What do you mean by primary?

ORÓ: That they are not the products of regeneration from sediments, and that they reflect the status of the early material on earth.

SIEVER: Impossible! You are assuming that carbonatites are the original old materials, and they are not.

ORÓ: I'm not assuming anything, but only arguing what authors in a book of carbonatites have said. This apparently is related to some of the diamonds found in the kimberlite as being magmatic intrusions.

I am not a geologist; this is why I ask. I'm only referring to Harmon Craig's measurements of diamonds in kimberlite columns. He has obtained δC^{13} of -7. This is the only terrestrial carbon that he would consider primary.

I don't know whether it is fortuitous (it doesn't make good sense) but the measurements on C^{13} available for the carbonatites are at the same value, -7. The C^{13}'s vary between about -6 and -9.

SAGAN: From the fact that I have never heard of a carbonatite, I deduce they are not very common.

ORÓ: They are very common and there are many very common things that you have not heard about!

SAGAN: That's possible. But certainly no significant fraction (say ten per cent or more) of the carbon in the sedimentary column can be in carbonatites.

ORÓ: I am not saying that.

BARGHOORN: Are you regarding the kimberlite types as carbonatites?

ORÓ: I don't know. This is why I used two different names; if you tell me they are the same, fine.

BARGHOORN: The kimberlite types came up, apparently, during the Jurassic.

SAGAN: Are other forms of carbon found in the early Precambrian other than organic carbon?

SIEVER: Are you trying to get rid of organic carbon by oxidizing it to carbonate in the ocean?

SAGAN: No. I'm saying: Here is a pigeonhole for abiotic inorganic carbon, here is another and here is another. That's all the pigeonholes there are. Everything else must have been organic. How much is that everything else?

BADA: But one thing must be remembered. There are no rocks for the first billion years of the earth's history. At 3.5 billion, in our oldest rocks — we know that life was on its way, there are microfossils. So there is no geological evidence for this.

SAGAN: There may be no evidence whatever!

BARGHOORN: Would somebody like to speak on the stability of such significant organic compounds as tetrapyrroles and prophyrins?

ABELSON: I have studied them to some extent. There are interesting and more likely stable organic entities to look at than amino acids. There are also improved ways of looking at optical activity. The amount of sterols is not great, but once a sterol is formed it can be reduced and incorporated in a sedimentary rock, and its optical activity remains. The organic geochemical search for steranes is one of the better things to do.

Tom Hoering (Carnegie Institution of Washington) studied sterane from petroleum. We know that petroleum itself does not form unless the temperatures have been up to 130°C over quite a few years – some millions of years. This is testimony with respect to the porphyrins. The fact of a substantial amount of porphyrins and steranes in petroleum indicates these structures persist under pretty severe geological conditions. If they tolerate 130-150° for some millions of years, they ought to be preserved somewhere where they can be studied.

We also isolated fatty acids from at least two different Precambrian formations. One was a stripey marker associated with the Nonesuch shale and another was from down in Australia (Hoering, 1967). In both instances the amount of organic matter was very substantial, distributed uniformly laterally through the rock. It was a situation where you really had to strain to assume that this stuff had been brought in.

In a limited number of formations, the amount of organic matter is so great that it becomes impractical to assume trickling groundwater has produced a large deposition of organics over quite an area.

MILLER: Isn't oil accumulated by means of trickling groundwater –

ABELSON: Who knows how oil is accumulated? But if anything is known for sure it is that oil migrates. But I was speaking of fatty acids, and the content associated with a certain sediment formations.

BARGHOORN: Oil migrates up, but not down.

ABELSON: That is something you have going for you.

SAGAN: I am still puzzled by Jeff's comments. The oldest rocks are only 3.5 billion years old and since we haven't evidence from the first billion years of earth history, it is of no relevance to determine what is the form of carbon in the oldest sediments we have.

If, for example, carbonates are genuinely very rare 3.5 billions years ago and there are two sources of carbonates, one biogenic, and the other abiogenic, the maximum abiogenic production 3.5 billion years ago is the observed amount.

If we could add up sources of 3 or 3.5 billion year old carbon, and found that a very small fraction of it was inorganic, could we not deduce that a large fraction of the remainder was organic and that the abiotic organic concentrations were large?

Is there something wrong with the argument?

SIEVER: Yes. It can't be done.

SAGAN: Why not?

SIEVER: Because we have no data. We have got one rock –

SAGAN: Let us take 2.5 to 3 billion years ago, then.

SIEVER: Let us take something 400 million years ago, which is a lot easier, and still it is terribly difficult to estimate the relative amounts and source of carbonate.

SAGAN: Is your point that the data are not good enough?

SIEVER: We have orders of magnitude.

SAGAN: What are they?

SIEVER: Let us do it for the present. The ratio is about 2500 carbonate to around a 1000 buried organic compared to 1 for the total of atmospheric CO_2. It is about that.

ABELSON: Yes.

SIEVER: It is about 2.5 to 1, carbonate to reduced carbon. That is after life was well started.

SAGAN: Part of that organic is of biological origin. Probably most organic material is biogenic, so this doesn't really help.

SIEVER: It is hard to conceive of any of it not being biological.

SAGAN: O.K. It can't be done yet.

NAGYVARY: I have an unrelated question. Could someone consider olefins in prebiotic speculations? I would like your opinion on the stability or possible occurrence of olefins in prebiotic media on the ocean or wherever.

ORÓ: Leslie can probably talk about cyanoacetylene, which is a kind of olefin. Perhaps Cyril is prepared to talk about the possible olefins in the Murchison meteorite.

NAGYVARY: I mean plain olefins, like isobutene.

MILLER: Have you some reason for interest in isobutene?

NAGYVARY: Yes. Very good reasons. Some of the olefins can add very rapidly to thiophosphates, and lead eventually to phosphates. This could give an easy type of activation, and also protection, which is used in organic synthesis. It might have some interest in prebiotic synthesis.

MILLER: Is it a radical or a neutrophilic addition?

NAGYVARY: No, it is a straight acid catalysis.

MILLER: All sorts of unsaturated molecules can be made very nicely by electric discharges. They are also made in good yield in pyrolysis type reactions. The quantity of these in the atmosphere is another matter.

NAGYVARY: Do you think they might be stable on the surface of the ocean?

MILLER: They don't react — but isobutene is pretty volatile.

PONNAMPERUMA: They also saturate pretty fast.

NAGYVARY: In the absence of a catalyst? By hydrogen, photochemically?

PONNAMPERUMA: The surfaces might be the surfaces that could catalyze the reaction. I have no experimental evidence, but this is suggested from the extreme reactivity of some of these molecules.

NAGYVARY: You expect cyanoacetylene to be very reactive and to react very fast after its formation, right?

ORGEL: Yes.

LEMMON: These can just as easily polymerize to become saturated hydrocarbons.

MILLER: Then where will the acid come from for acid catalysis?

NAGYVARY: Isobutene needs an acid catalyst, but other types, like acrylonitrile don't need anything.

MILLER: A nitrile is different. That reacts with ammonia to make the nitrile of β-alanine, for example.

NAGYVARY: But thiophosphates would react much faster than ammonia.

SCHOPF: I thought it might be useful to briefly pursue something that Cyril raised. Is it worth looking for prebiotic kerogen in very ancient sediments? How could such kerogens be identified and distinguished from biological kerogens?

Nagy's report in *Nature* of the occurrence of highly aromatic material in pyrolysis mass spectrometric analyses of the lower Onwerwacht kerogen comes to mind (Scott, *et al.,* 1970). I wonder if this fits anyone's definition of what might be expected for prebiotic organics.

PONNAMPERUMA: In the very same rock they have found the kind of microfossils you have seen in the Fig Tree.

SCHOPF: That is a different question. The source material for their pyrolysis mass spec experiments came from near the base of the Onwerwacht; their samples occur substantially lower in the column than the Fig Tree material.

It is an interesting organic geochemical report and I'd like to know what you professionals say.

ABELSON: I will answer as a professional! Carbon compounds tend to go to CO_2, methane, or graphite. They tend to make polyaromatic rings. That is doing what comes naturally.

I don't think the finding of polyaromatic compounds is all that significant. However, if we have a prayer of finding prebiotic kerogen, it would need to be somehow via the R groups associated.

For example, in present day kerogen by a very partial oxidation and liberation of R groups, in principle you should get some evidence, at least a smell, of say, leucine. A prebiotic kerogen might well show up with simpler R groups — you might only be able to extract 2-carbon or 3-carbon compounds, a simpler sort of partial oxidation product.

BARGHOORN: Did they look for coronine?

SCHOPF: I don't know, but with regard to the aromatic argument they presented, metamorphism apparently can not be used as the explanation; they have done the same experiments on several other samples higher in the column, but from the same geologic terrain and subject to the same geologic history, and there they find more typical paraffinic material in pyrolysis mass spec. If that is so, I don't think the material has been metamorphosed or degraded.

MARGULIS: They got aromatics in both the Onwerwacht and the Fig Tree. It was just the absence of typical straight chain material in the more ancient sediments that was significant.

ABELSON: One must be careful. Tom Hoering found straight chain hydrocarbons in one of these South African-sediments, the Fig Tree, and got enough to do isotope analysis on it (Hoering, 1967). It turned out that the isotope fractionation on the hydrocarbon was considerably different from that on the insoluble portion. This is highly suggestive of the migration of hydrocarbons.

PONNAMPERUMA: Alkanes will fall into the same category as amino acids because of migration.

ABELSON: Suppose one part per million or so is pyrolized off the carbon of the shale. This is just in the range of adsorption. You need a richer source.

SCHOPF: Let me ask a different question. Would anyone speculate as to what carbon isotopic values of abiotic kerogen might be in comparison with younger biotic kerogens? Might that be used as a tracer?

ORÓ: This is another controversy. Nothing we know today allows any good answer. Even comets have a C^{13}/C^{12} ratio comparable to the terrestrial one.

SCHOPF: But do Type I carbonaceous chondrites and carbonatites?

ORÓ: On the average most bound carbon in carbonaceous chondrites has a value of about -7 as I said. As far as I know all of the graphite in iron meteorites that has been measured comes to the average value of -7. The carbonatites also have the value of -7. Whether this implies a common source I don't know.

SCHOPF: But lunar carbon is somewhat different?

ORÓ: Lunar isotopic carbon values are slightly on the positive side: roughly up to $\delta^{13}C \cong +20$ per mil.

There are other reported negative values on lunar material, particularly for igneous rocks. However the existence of these negative values may be related to the problem of contamination that I discussed Wednesday. An igneous rock with a very low carbon content has a high probability of its carbon isotopic value being made negative by a small amount of terrestrial organic contamination. Therefore, most of the C^{13}/C^{12} values reported for igneous rocks may be suspect.

Now, we may attempt to relate the carbon isotope fractionation values of our hypothetical abiotic kerogen to other values in the solar system. Aside from meteorites, I think there are one, two or three scanty values reported from comets. On the average these values come out very comparable to the values for terrestrial organic compounds.

SCHOPF: -25?

ORÓ: Yes.* So even though carbon isotope fractionation ratio is a good tool to analyze different fractions of a sample on the earth to distinguish inorganic carbonate and organic carbon, it is not good enough to go back to the Precambrian and then hope to find a significantly different value there in abiotic material.

SCHOPF: It is still conceivable that one might find a significantly different value in early terrestrial abiotic kerogen − one that would be closer to zero, or to -7 or -15 than say, to -25 to -30.

BARGHOORN: It's time for lunch, and I would like to say a last word, if I may!

There are new dates for the lower Onwerwacht of south Africa published last year: these oldest sediments may be as old as 3.36 billion years. The dates for the middle of the Fig Tree is at 3.2 billion. If there is such a difference chemically, it means the origin of life can be pinpointed to have occurred during a pretty small fraction of time.

---

*Actually the values obtained by optical measurement of the cometary $^{12}C/^{13}C$ Swan bands don't have the accuracy of isotopic ratio mass spectrometric measurements. They are measured in percent instead of per mil, e.g.: Comet Ikeya $\simeq 70 \pm 15\%$ (Stawitowski and Greenstein, 1964); Comet Tago-Sato-Kosaka, 1969g, $\simeq 90 \pm 15\%$ (Owen, 1971). The only conclusion possible is that they are similar to average terrestrial values.

SIEVER: I first would like to discuss some current thinking about the evolution of the early atmosphere and the general geochemical cycle on the surface of the earth. Ideas of today differ from those of four years ago, when I was last at this origin of life conference, because of a rather profound change in geological thinking: plate tectonics, which many of you know about.

LEMMON: Some of us don't.

SIEVER: I'll talk about it.

First, a consideration of plate tectonics will set the stage for a background of the large scale general dynamics of the earth's surface. Second, I want to discuss my diagram (Fig. 14), relating the question of dynamics to how production of primitive small molecules led to rather large ones, in particular, what kind of transportation processes could be invoked and where. Third, Carl Sagan wants to talk about the even earlier thermal history of the earth, which I want to mention also.

I will start with a now very popular idea that William W. Rubey first proposed in the early '50's. Rubey re-examined and quantified some older ideas in a very beautiful way. He said simply: atmospheric weathering of igneous rocks, such as granites, forms clays, sands, and limestones, salts, and so forth. This was a process for which a mass balance sheet could be made. Add so much CO_2 and water to an average igneous rock, and take away so much sodium and potassium, and so on, and the average sediment could be made from the average igneous rock.

If all sediments known on the surface of the earth from geology were mapped and the original amount of igneous rocks needed to produce them calculated, a nice balance could *not* be made. In fact, there was an excess of certain materials needed to produce sediments – "excess volatiles". They are all gases: CO_2, water, hydrogen sulfide, HCl, much of which reacted with surface rocks and are incorporated in them. For example, most CO_2 that has come from deep in the earth is now buried as limestone and the organic carbon deposits, oil and coal.

If the total amount of CO_2 currently in the atmosphere is taken as 1, there are about a few times the total amount of CO_2 in the atmosphere in the surface layers of the ocean. About sixty times that are in the deep parts of the ocean, and 1000 times that in the form of reduced carbon in the sedimentary crust. About 2500 times the amount of CO_2 in the atmosphere is in limestones. These figures have been revised slightly since Rubey's work but the general orders of magnitude are correct.

That is far too much CO_2 than can be accounted for by the amount carried in igneous rocks. Granites simply don't contain enough CO_2, HCl and other materials. Gases coming out of volcanoes today contain largely water and smaller amounts of CO_2. Occasionally measurements of other gases: HCl, methane, nitrogen,and ammonia, and several other gases have been made. It is difficult to say what the "average" volcanic emanation is. But Rubey simply proceeded from measurements of volcanic gases as the skeleton of his logic to deduce that most of these excess volatiles were gases that must have outgassed from the interior of the earth earlier. Since hydrogen (and helium at a slower rate) are the only gases constantly lost from the earth's surface, none of these other gases ever reach escape velocity and leave. Therefore, the excess volatiles must have accumulated during earlier geologic time.

He then asked: Did all atmospheric gas come out at once, did it come out gradually, or in some other odd way? He reasoned as follows: if these excess volatiles had all come out at once, we would have had a very dense atmosphere of CO_2, HCl, H_2S, or SO_2, depending upon the amount of hydrogen present at that time. (The hydrogen would influence the oxidation state of the gases). Not only would these large quantities of gas have made a dense atmosphere but also a very acidic ocean which would have been intolerable for any life. He estimated the pH would have been about 1, or thereabouts, and that, therefore, it was exceedingly unlikely that this has occurred.

He then argued on biological evidence: Most living forms known from paleontology are similar to modern forms suggesting that there was not a vast change in salinity tolerance during known geologic time, the last 600 million years. Therefore, the oceans probably have not changed drastically in pH or salinity during the Phanerozoic.

From this he reasoned the addition of all the gases to the atmosphere must have been very slow. Add up the chloride, the carbon dioxide, the sulfide — all these excess volatiles and simply divide by roughly four billion years, the flux is extremely small because the time is so long.

Since he rejected the idea of such an acidic ocean, he claimed the simplest assumption was a linear rate of degassing through geologic time. That is the Rubey model.

Since then (Rubey, 1953) aspects of that model have become rather clear. First, it is obvious that linear rates of degassing is just a first approximation.

Secondly, the fundamental basis of the assumption (that you can match up all sediments now on the earth and mappable with the total amount of igneous rock that was eroded) is in question because of global tectonics. Rubey

assumed all sediments are recycled only through the very surface parts of the
earth, and eventually will appear as metamorphic rocks. Outgassing was assumed to come from deep inside the interior of the earth, and then stay at the
surface.

Global Tectonics

Global tectonics has thrown a fishhook into that idea. Global tectonics
was first discovered in a study of the mid-Atlantic ridge. (Fig. 12).

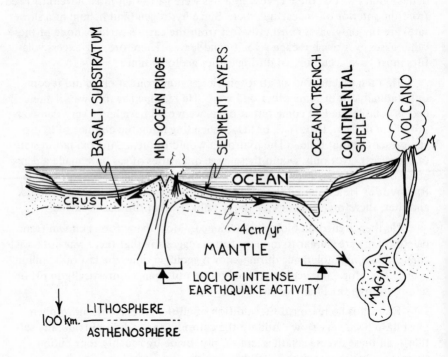

Figure 12. Plate tectonics and the cycling of the lithosphere.

Here is a cross section of a typical ocean (which doesn't really exist)
which shows a ridge in the middle of the ocean. This is where material apparently comes to the surface and spreads laterally. There are many differ-

ent lines of evidence; all converge to show the same lateral sea floor spreading. The plates of the lithosphere are about 100 km thick.

This must not be confused with the depth of the Mohorovicic discontinuity, which occurs very shallowly underneath the surface of the ocean. It is just a fraction of the thickness of lithosphere plates.

The evidence (in this very degenerate form that I am discussing it) is that material comes up where new lithospheric crust is being created and new lithosphere is made from what is called the asthenosphere, a place where behavior is nonelastic. This material moves up and out at ridges — and at certain other places, typically deep oceanic depressions, like the Mariana Trench in the Pacific, there is actually a corresponding downward movement. At these sites the real sharp evidence for the downward process came that completed the dynamic balance. Analysis of earthquakes foci showed that oceanic lithospheric plates were descending into the upper mantle at a low angle to the horizontal. The dynamics of these plates involves the production at one place, the ridge, and destruction at another, the trench. As the descending plates move down into the athenosphere, they start melting, behaving viscously. Then they get recycled into the upper part of the mantle. I won't discuss motions within the mantle, but they are heterogeneous and involve certain kinds of mixing. The depth of the mixing is guessed to be of the order of 300 to 400 km. This gives sort of a mixing length for the surface part of the earth which is far deeper than anybody had ever thought. The mixing length that was implied by Rubey for recycling the surface rocks of the earth was much less than 50 km. We have to consider a much deeper kind of recycling.

What happens to the sediments that are produced in oceans? Of course it is more complicated than my picture, but essentially sediments are produced, and a fraction are gradually funneling back down into the interior of the earth, down deep into the mantle. They are then metamorphosed, melted, completely lose their identity, and become mixed up with the viscous material which finds its way back up as igneous magma.

From this "machine", understanding this process, we think we understand the production of almost all of the known rocks. Carbonatites are a problem, and there are other problems, that I don't think I need to bore you with. Most importantly the idea of making a mass balance between the total amounts of sediments which we have left now on certain continents and the total igneous rock weathered in the history of the earth is not susceptible to solution the way Rubey tried it.

In most places sediments are systematically destroyed by erosion; the curve of the rocks preserved in the geologic column as a function of age is like that shown in Fig. 13 (Garrels, Mackenzie and Siever, 1971). There are no sediments from 4.7 billion years ago until about 3 billion years ago and very little more until more recent times, the last few hundred million years.

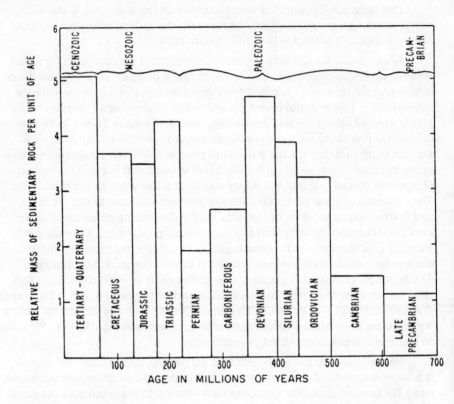

Figure 13. Relative mass of observed sedimentary rocks as a function of age. The area of each histogram block is proportional to the mass of rock of that period that is found today divided by the duration of the period.

Because of the constant destruction of sediments by erosion, plus the idea that at least the upper 400 km of the earth has been recycling, some of us have concluded that this behavior of the earth (probably ultimately a response to radioactive heating in the upper mantle) was initiated as soon as the earth differentiated into core, mantle, and crust. The machine simply started early, as soon as differentiation started.

From both meteorites and now from the moon we have evidence that differentiation could have taken place extremely early in the history of the solar system. Previously we considered longer time scales, approximately a billion years needed for the differentiation and the cycle to start.

Now we think that within a few million years the planet differentiated because of the very old rocks on the moon, as well as evidence that the meteorites came from a differentiated, exploded planet. The ages of the meteorites suggest that by 4.6 or 4.5 billion years the earth had differentiated and the cycle begun.

SAGAN: To the extent that differentiation occurs because of the gravitational potential energy of accretion, the earth has more gravitational potential energy per gram than either the moon or the parent bodies of the meteorites. So your argument is even stronger.

SIEVER: Right. By differentiation I am referring just to sedimenting the denser parts of the earth to form the iron core.

BARGHOORN: Ray, where does the curve start (Fig. 13)?

SIEVER: Approximately Churchill province, 2500 million years ago. There are so few rocks earlier that there is no abundance to them. Essentially this is what Rubey hadn't known about, that much more than the thin crust down to the Mohorovivic discontinuity is recycled; a mass balance simply can't be made.

We ought to rethink another Rubey thought: that a dense primitive atmosphere resulting in a very acidic ocean is unacceptable. Following Herman Kahn, sometimes I like to think about the unthinkable. Since Rubey first projected this we have discovered that weathering reactions are very fast. We used to think silicates react very slowly, and therefore weathering was slow. Now we can follow these reactions in the laboratory; they really are extraordinarily rapid.

When HCl, CO_2, H_2S come out of the earth, these gases immediately hydrolyze and attack the rocks, resulting in formations such as limestone and clay and various alkaline metals in solution. Then alkaline earths appear in solution, if they are not soon precipitated out.

What would happen if there were, early in the history of the planet, a large, rapid outpouring of gases from the interior, one that even may have been associated with the fast differentiation of the earth? Essentially and simply, it would have been comparable to any acid-base titration. It doesn't really make much difference whether you pour a gallon of HCl on an outcrop of limestone or add it a drop at a time. CO_2 will simply come off faster. I think what really happened is that there never was a dense primitive atmosphere, because as soon as these gases came out, the rate of weathering simply increased with the flux of outgassing, so that the early primitive crust was probably limestone-clay. This makes me very happy for the origin of life, and I think should make all of you very happy for the same reason.

In other words, I see no compelling reasons against an early, large-scale outgassing, although it is still entirely possible that there was a linear rate of degassing, there is no particular way to do it and it would imply that the oceans grew through geologic time. Since with the linear rate of outpouring of H_2O the total amount of water coming out gradually accreted the oceans, 3 billion years ago they would have been a fourth as large in volume of water as there are now.

MILLER: What is wrong with that idea?

SIEVER: Only one thing. Some recent hypotheses on the evolution of the continents which depend on evidence including certain arguments on strontium isotopes, suggest that the continents have always been here though not necessarily of the same size, thickness, or distribution. If the continents have always existed in roughly their general volume, the relative heights of oceans and continents must have been approximately what they are today. The adjustments in density of the two different kinds of crust implies that the oceans were much smaller; they must have been a little puddle at the bottom of deep ocean basins, yet we know from geology that at least as long ago as two billion years there was extensive flooding of continental areas by the oceans, which means sea level must have been close to the margins of the continents.

So the logical chain based on the arguments for continent evolution which more and more people are beginning to accept, is that if the ocean basins and the continents were much as they are today, though with a different distribution over the face of the globe, the ocean basin must have been filled with water.

MILLER: Of course others have said the opposite with such confidence, namely that the oceans were smaller four billion years ago than they are now.

SIEVER: Naturally, I give you what I believe first. For example, Al Engel (University of California, San Diego) does not agree.

SAGAN: How can the earth have differentiated very early without a great deal of early outgassing?

SIEVER: I agree that is an eminently reasonable point of view!

SAGAN: It is a very straightforward conclusion.

SIEVER: Actually, we don't know too much. The rates can only be inferred because there are no rocks older than a little over 3 billion years left.

SAGAN: But the deduced concentration of radioactive thorium, potassium, and uranium in the crust of the earth had to have occurred when the earth differentiated.

SIEVER: Right.

SAGAN: And that is understood by V. M. Goldschmidt's old argument that the earth was molten and cools; the ionic radii of these radionuclides are too large to fit into the silicate crystal lattice, and thus are sweated up.

If they sweated up 4.5 billion years ago, a great deal of occluded gas must have come out at that same time. If the earth differentiated early, there must have been massive atmosphere. I don't see how it is possible to escape that conclusion.

SIEVER: I don't either.

SAGAN: What does Engel say?

SIEVER: I think that Engel thinks the continents have actually accreted since the beginning of geologic time. He believes the continents started out as very thin, relatively smaller plates; that the ocean basins were differentiated with respect to the continents, but because the continents were so much smaller, the ocean basins were not that much different from them in depth. The ocean basins were smaller, and therefore the amount of water in them was smaller. And then as the continents accreted, they became thicker and larger, and started popping up higher, with the creation of a denser crust underneath the ocean, giving you a deeper ocean basin. This means the rate of growth of continents parallels the rate of outpouring of water which fills up the ocean basin. It is a very neat machine.

BARGHOORN: Perhaps this is a technicality inappropriate here, but isn't your model really based on the north Atlantic, south Atlantic history.

SIEVER: No, it's based on the world ocean.

BARGHOORN: Okay, you have a cause-and-effect relationship, the drifting of the continents as concomitant, but how about the basalts? Do you accept the process of plate tectonics throughout geologic time? Was there drifting throughout the total record, or was it just post-Paleozoic drift? In other words was there predrift drift? This is fundamental.

SIEVER: The model says what the dynamics are right now.

BARGHOORN: Right.

SIEVER: The heat that is driving this engine is radioactive decay. When we project backwards in time there is no particular reason why we should substitute another engine, or no engine at all, for this. In fact, we probably had more heat, as we had more radioactive decay in the past. Therefore, I conclude we probably had more plates and more tectonic activity earlier, particularly at the very early time, when a lot of radioactive heat was coming out.

SAGAN: But one difficulty is the reconstruction of the Gondwanaland?

SIEVER: Yes. That's so recent.

SAGAN: I have never understood something related exactly to what Elso said: is not one of the major justifications for the whole plate tectonics argument continental drift?

SIEVER: No, that's a misconception. Continental drift just sort of falls out of the side pocket of global tectonics. Global tectonics as an engine is independent of continental drift.

SAGAN: Okay. But wasn't there a time quite recently (certainly in Phanerozoic times) in which all the continents were together in one continental mass?

SIEVER: Not necessarily.

SAGAN: That is the usual reconstruction that one sees.

BARGHOORN: One or two continental masses.

SAGAN: Yes, one or two.

SIEVER: That may have been transitory. For example, one current reinterpretation of the North American continent versus South America and Africa is that they were once apart, they came together in the early Paleozoic, and then split apart again. The Mediterranean is a site where Africa and Europe are coming together now. The Himalayas are another site where two continents have come together very recently.

MARGULIS: Lake Baikal, USSR, is a place where they are separating, so presumably it has been going on. Essentially we are sampling at different times to see the drift.

SIEVER: That is right.

HULETT: I thought that the heat down below a few hundred kilometers really did not have time to escape, so there really was no more heat in the early history of the earth. We essentially have all the heat that was there early, plus what has been generated since then. Then I thought it was concentrated in the top 20 or 30 km. How does that coincide with the fact that (at least under the ocean) the top 20 or 30 km seemed to be going back down underneath all the time, which would seem to prevent the concentration required to give us the heat we seem to have now?

SIEVER: There is a large heat flux from the interior over the whole surface of the sphere, so the earth is not cooling, but there is a steady heat loss from the interior. It is not being bottled up and the rate of heat production and loss is still more or less in a steady state over the time scales we are analyzing.

HULETT: What about the recycling of the surface?

SIEVER: The recycling is asymmetric: the ocean basins recycle much faster than the continents. The continents tend to float and when two continents meet a slippage tends to occur so that the less dense granitically sedimentary parts of the crust tend to pile up in mountain chains and only the deeper basaltic parts of the crust tend to get sucked down. This is more or less the general idea.

It is an asymmetry, not an all-or-none proposition. Obviously, much sedimentary and granitic crust also gets drawn down too, but it gets entrained, you might say. You may be interested that the rate is now usually given at about 4 cm per year. The satellite geodesy program, now in progress, is supposed to actually measure the separation of the Atlantic ocean, if it can be measured.

BADA: They have measured beautifully the rate of separation of the two plates in California from the recent earthquake. On the San Andreas fault one plate is moving south, and the other north, and it moved 4 or 5 cm in just one big shock (Anderson, 1971).

ABELSON: Ray, a picture you were developing before these side issues started was related to the minute the volatiles came out, quickly giving you weathering, corrosion and neutralization. This would depend on the presence of land relief. If the whole volume of water of the ocean were averaged out over the whole surface of the globe, there would be − I forget − something like 13,000 ft. of water: If erosion occurred then somehow there must have been early tremendous relief. If everything is under water, there couldn't have been much weathering.

SIEVER: That is not true, because we can hypothesize volcanoes under the ocean, thus volcanic emanations (CO_2, some HCl) come out underneath the oceans and simply do their weathering work on the way up.

Now obviously, volcanic emanations in the present ocean are diluted terribly, but there is another strand of evidence for these ideas. A few years ago a study was made of the Yellowstone hot springs area where there is an upward motion of HCl. The HCl never reaches the surface, because it weathers rock on the way up. This poor HCl doesn't know it is supposed to get up to the surface before it starts to weather, and as it cools down on its way up through the rock, due to the porosity of the rock, it starts the weathering process which accounts for the origin of a lot of mineral deposits. It is called wall-rock alteration. I think this happened: the stronger the acid, the more likely it would have reacted to weather material before it even got to the surface. The reason we see mostly CO_2 in the air, rather than HCl is because HCl never got a chance.

It is entertaining to add up the total rate of proportions of CO_2 to HCl that ever emanated from volcanoes. This could be summed up if the record was complete of the total amount of chloride in the earth versus the total amount of CO_2. We used to think it was about three to one in favor of CO_2. More recently, it is thought to be more like one to one, implying an enormous amount of HCl has come out. What happens is the production of evaporites. The sodium chloride from the evaporation of sea water usually is deposited at the edge of the sea someplace. These are peculiarly susceptible places to get recycling, and presumably much chloride has been recycled.

MILLER: What happens to the sodium when HCl is made from the NaCl?

SIEVER: There is a whole countercycle which occurs deep down, is that what you mean?

MILLER: Yes.

SIEVER: That is subject to something else I want to discuss later: how the total amount of oxygen in the present atmosphere does not have to be tied to photosynthesis, because, in fact, it could have been used in the oxidation of reduced carbon deep inside the interior. Reduced carbon brought into the interior, can be reacted with, for example, ferric oxide, which is also being brought down. This nice couple can produce then CO_2 plus ferrous iron. In fact, all the deep-seated rocks coming up from volcanoes contain reduced iron; and yet from the cycling process we know oxidized iron goes down.

Therefore, the argument of Wally Broecker (1970) on the dynamics of oxygen that there can be only as much oxygen in the air as reduced carbon buried is based on the incorrect assumption that the reduced carbon gets buried, whereas all the oxygen does not. You see, photosynthesis produces a mole of (CH_2O_n) per mole of O_2. Most of the oxygen stays in the atmosphere except to the extent that it gets buried as ferric oxide. There are also nitrogen compounds that can be reduced, that is, N_2 can be reduced to ammonia.

MILLER: Is that going on?

SIEVER: No, it is all hypothesis.

MILLER: I realize it is hard to get data but do you think it is going on?

SIEVER: It must. If you just think about the possible redox couples in the interior of the earth, what do you have in terms of the abundant elements to play with? Not very much.

SCHOPF: Nevertheless there does seem to be a reasonably good match between buried carbon and O_2 if Holland's numbers are used, suggesting that the process you refer to may not be quantitatively important.

SIEVER: It all depends on the numbers which I haven't really carefully investigated. But it says that if all the reduced carbon — oil, coal, everything else — at the earth's surface is burned, the last mole of reduced carbon and the last mole of oxygen in the air will be used up to make CO_2 and H_2O.

SCHOPF: Not just in the air, but in the oxygen sinks as well. In both Holland's calculations and those based on Rubey's estimates the numbers are within an order of magnitude.

FERRIS: But this article in *Science* (Van Valen, 1971) says they don't agree, which is the dilemma that someone like myself is in.

SIEVER: That's right.

FERRIS: Van Valen is very convincing that it just can't be.

SIEVER: Some numbers are wrong. That is the problem, when you start estimating orders of magnitude, Carl!

Sedimented Organic Carbon and Limestone

BARGHOORN: But most reduced carbon in the earth (coal, oil, gas, and organic matter in living material) is peanuts compared to the reduced carbon in black shale. Isn't that where the great bulk of it is? Rubey's figures are very high. Carbon in coal, oil, and all living organic matter is 28 x 10^{18}; limestones are 62,500 x 10^{18}; and buried organic matter is 22,000; the great bulk of the carbon is locked up in these sedimentary rocks.

SIEVER: I can't go any further, because I haven't really calculated these for myself, but the point of what I am saying is that a reasonable setting to begin producing life is on an earth not very dissimilar from the present one in some respects. Yet I think the earth would have been recycling then at a more rapid rate, because of slightly increased heat sources. From considerations of prebiotic syntheses I think there was no free oxygen around. I can't evaluate myself what the rate of photolysis of water is in the upper atmosphere, and whether it is an adequate source for oxygen, rather than photosynthesis; but it doesn't seem very likely to me that today UV photolysis is a very important source of oxygen.

SAGAN: Not today; but an ocean of water can be photodissociated in geological time, if conditions are right.

SIEVER: There continue to be arguments for the importance of photodissociation in the upper atmosphere, so that subject is still alive, but I cannot evaluate the arguments.

BADA: If the ocean were photodissociated, how much oxygen would it yield?

SAGAN: Vastly more oxygen than in the atmosphere, and if it were to be removed from the atmosphere, something highly reducing in the crust would have to grab it up, because it surely couldn't escape from our gravitational field.

BADA: How much oxygen from carbonate?

SAGAN: There is about 300,000 gm per square centimeter of ocean, and every molecule of water is mostly oxygen. So you have 300,000 gm per square centimeter of oxygen. How much oxygen is there in the atmosphere? About 200 gm per square centimeter of oxygen. And how much oxygen is there in the crust? At most, the CO_2 content of the sedimentary column is about 100 atmospheres. Most of the oxygen obviously is in the ocean.

SIEVER: But this doesn't really speak to primitive conditions. The present oxygen level may have come either from photolysis in the upper atmosphere or photosynthesis − I'll leave that as a moot point, though I believe it is photosynthesis, but really I have no basis other than prebiotic synthesis for saying that there was no very great amount of oxygen present early. I think it is fairly probable that oxygen was low but I have no reasons besides prebiotic synthesis − perhaps others have.

MILLER: The standard argument given by Urey is that the carbon was all reduced in the original gases because of the abundance of hydrogen in the solar system and the universe. Elements that were then reduced are now oxidized, including carbon, nitrogen, and the 230 gm/cm^2 of oxygen in the atmosphere, as well as the iron and sulfur. This implies several thousand grams of water, the equivalent of which was photodissociated, and the hydrogen escaped. That is the standard argument. But with the present cold trap, this can't have occurred; enough water to make even the 230 gm of oxygen in the atmosphere can not get through the present cold trap of the tropopause in two billion years. I think that methane will also go through the cold trap without difficulty. If you change this so the limiting factor is not the cold trap, but rather the rate of diffusion to the upper atmosphere the situation is greatly changed.

SAGAN: Since the average temperatures were less, I suspect the cold trap was even more important than now, and therefore the rate of water diffusion to the level of solar dissociation is even a stronger problem.

MILLER: I have always been struck by the fact that the hottest part of the earth — namely, the equator — has the coldest cold trap.

SAGAN: That is just a circulation problem.

MILLER: It would become dominant in your consideration of the temperature of the cold trap in certain regions.

SAGAN: The average temperature of the earth was down by some tens of degrees in early times, and so the cold trap must have been affected (see p. 205).

To first order, the stratospheric temperature has a constant proportionality to the surface temperature. One of the simplest radiative transfer solutions for this factor is $2^{-\frac{1}{4}}$.

ABELSON: Will we hear from you on this later?

SAGAN: Yes.

ABELSON: Several facts are possibly at variance with your picture. One is that really there isn't very much calcium carbonate in the earliest rocks. There's a little bit of it, but not very much. And furthermore there isn't very much reduced carbon which would have been expected if there had been a tremendous reducing situation with available material to be somehow polymerized and drained down into the sediments.

How do you interpret the fact that, if anything, the oldest rocks look like there were less volatiles around?

SIEVER: I think there is a selective preservation of rocks. Limestone is the most easily weathered rock. Think of most cycling elements as a collection of flywheels, some going fast and some slow, the limestone was probably very rapid and accounts currently for most of the weathering.

ABELSON: This is all right where the limestone is off by itself. But what is your explanation where it is part of a general geologic column, where sediments are above and below, and the limestone hasn't been too exposed?

SIEVER: Let me refer to the most interesting recent studies of Paul Hoffman, largely in a limestone terrain in the Great Slave Lake area of Canada. These have been ignored. Some of those areas have not been mapped, implying it is probably a selective mapping as well as a selective preservation effect. I can't prove it and I agree that this is a difficulty.

SCHOPF: Al Engel, who has studied the Swaziland supergroup rather extensively believes there are large amounts of limestone in it. He is also thinks the Buluwayan limestones, rocks 3.4 to 2.8 billion years old, are extensive. There are few areas on the earth's surface that have such ancient rocks and even fewer have been studied in detail. Here in South Africa is a good example, and according to Engel who has worked there, substantial amounts of limestone were deposited.

My impression which could be erroneous, is that the old saw about there being substantially less limestone at earlier times has not been firmly established.

Considering the amount of organic carbon, there clearly is some organic carbon in these early sediments. I don't think there is an either increasing or decreasing trend as you go back through time. Rather there is sort of an average quantity of limestone deposited; it might be found in the lower Phanerozoic or in the late Precambrian, when things were pretty much as they are today.

MILLER: It was once a tiny quarry. What is Al Engel's evidence that there is a great deal of limestone in the Buluwayan?

BARGHOORN: There is so little Buluwayan limestone that they abandoned the quarry when they removed what was there.

SCHOPF: That's not true! Limestones have been mapped in that general area, and correlated with the Buluwayan. The stromatolites occur in one small area where there is a limestone quarry, but there are limestones mapped throughout quite a lot of that area.

BARGHOORN: But agricultural limestone is so scarce in Rhodesia they have to import it.

SCHOPF: Except for a new date which hasn't yet been published, dates about 2.6 billion years old on mineral ages are based on pegmatites. There are about seven different dates, and they cut limestones that are correlated with the ones at −

BARGHOORN: Huntsman Quarry.

SCHOPF: One is 190 km; another 260 and another is 300 km; outcrops of the same limestone in that same area. Evidently it's not just a little pocket.

SIEVER: I don't think the statistics are good enough. It is just so small a volume of these rocks which has been so poorly mapped that I don't think we have very good estimates.

SCHOPF: I think there is a psychological fact that ought to be thrown into the equation: the traditional feeling about a decrease in limestone as one goes back in time. This may have colored people's views.

ABELSON: Maybe, but some pretty smart fellows were walking those outcrops sixty or eighty years ago. They were smart enough to know limestone when they saw it!

BARGHOORN: Limestone is a damned important mineral.

ABELSON: Some of those guys were a lot sharper than some of us around this table. They didn't swallow a tradition. They looked at the facts pretty carefully.

SIEVER: I have to disagree, Phil. In older reports you can see they did not proceed in the way that people who were measuring younger rocks did. They did not make a systematic assay of the total column. In the best of the older reports, say Van Hise and Leith, and others, half the time they did not recognize limestone. We know this from more recent work in the areas they mapped. Even great men sometimes can be preoccupied with other things.

SCHOPF: Prior to the last few decades they had no radiometric controls, too.

SIEVER: Certainly, for example, Van Hise called sediments late Precambrian or Cambrian that we now know were not.

BARGHOORN: Our argument here is based on a state of ignorance, yet in the Hammersley mountain range of northwest Australia there are 4,000 or 5,000 feet of siliceous rocks with banded iron, without a bit of limestone. There are a few dolomites. This can be seen for hundreds of miles; it is not just a small area.

SIEVER: We really need good statistics; without them there is no resolution to this argument right now.

BARGHOORN: I agree.

ABELSON: Okay.

Weathering

SIEVER: To continue, I think the general character of weathering and the production of sediments was the same then as now if the Precambrian pressure of carbon dioxide in the atmosphere was approximately the same or possibly a little higher than it is today. I estimated approximately 10^{-3} instead of

5×10^{-3}. Because most estimated weathering processes have little to do with redox potential, the whole question of oxygen can be left aside.

The hydrolysis of anhydrous silicates takes place regardless of whether the metals are reduced or oxidized. The CO_2/CH_4 ratio only depends on the rate of escape of hydrogen, after all. The production of clays and limestones with this quantity of atmospheric CO_2 should be proceeding pretty much as normal.

Another point: it has been learned fairly well in the last few years that the higher the mountains, the more intense the deformation, the greater the relief, the greater the amount of chemical weathering as well as the amount of physical weathering. Those of you who may have known the older dogma may consider this paradoxical, but, in fact, the more the mechanical weathering, the more the fragmentation of rock the more feldspar becomes clay, the various other kinds of silicates get dissolved. This has been beautifully demonstrated by a study of the High Andes and the Amazon River (Gibbs, 1967). There are now a number of confirming studies from various other places in the world.

The deformational state of the earth's crust or its relief may have been greater at earlier times, simply because the heat engine was going a little faster than it is now. This also implies the production of a great deal more sediment, which fits in with a more dense atmosphere, although I don't really have to invoke a dense atmosphere.

Leaving aside the question of oxygen and hydrogen in the atmosphere, sedimentation and normal weathering processes would have been going on much as we see them today in kind, although perhaps not in degree. This is the point.

BADA: You are saying that the buffering mechanism in the primitive ocean is exactly the same as the buffering mechanism in the present ocean.

SIEVER: That's right.

BADA: Ions like sodium, potassium, and hydrogen are therefore monitored by the clay.

SIEVER: Yes. The ocean is a place where the whole titration back reacts: CO_2 reacts with silicates to form clays and then the free bicarbonate which is produced is balanced by calcium, magnesium, or sodium, or some other metal which accumulates in the ocean. Carbonic acid plus silicate gives alkali metal exchange. Then in the ocean it back reacts, to give chemical sedimentation products and the clays, both detrital and disgenetic, act as a pretty reasonable buffer system for the ocean.

Nobody has been able to really knock down Sillen's (1961) argument. That is, the sheer abundance of the clays and their buffering capacity far outweighs carbonate or any other buffer system in controlling the pH of the ocean.

This brings us to my diagram (Fig. 14) that is based on the assumption that the early Precambrian earth was something like the earth today. This is a rough survey of the environments of the earth with respect to where molecules and ions are formed and moved from place to place. I have presumed, based on discussions with you, that pH, activity of water, and temperature are important with respect to prebiotic processes.

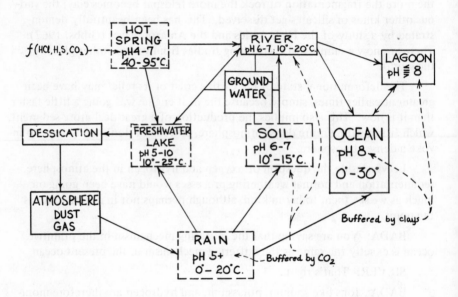

Figure 14. Surface environments of the primitive earth.

Let us start with the atmosphere, which would contain gas and dust; obviously its composition (although in some doubt) will depend on the time course of hydrogen escape.

Solid boxes refer to rather long residence times, and dashed boxes to rather short residence times. This estimation is not quantitative but in general probably of the order of a few thousands of years separates "long" from "short."

Molecules get washed out of the atmosphere by rain, and then dropped into lakes, soils, and of course the ocean as precipitation falls all over the earth.

The question of fresh water ponds or lakes as potential settings for the origin of life has been commonly mentioned. We can ask first of all what would be the pH of lakes. At the present time it seems to be a function of evaporation. That is, alkaline lakes tend to develop where there is relatively rapid evaporation, sometimes to desiccation. Even if not to desiccation, rapid evaporation in the average kind of water accumulating in a fresh water lake leads first to precipitation of calcium carbonate and depending on what ions come in, also to calcium sulfate and gypsum. The result is a concentration of strong bases left in the water and a rise of pH up to nine or ten.

There are also acid lakes, depending on the surrounding rocks. Acid lakes today have to do with drainage of acids from weathering, biological products, many associated with swamps — or people: there are acid mine waters which drain into acid lakes, and so forth. I rather doubt that in early Precambrian time when there was no biology that there would be extensive acid lakes except those fed by acid hot springs related to volcanism.

Most lakes probably fall into much the same relatively basic pH range as the ocean, if there is no rapid desiccation. Desiccation will probably drive the pH up.

I have given wild guesses for the temperature. I have assumed that the earth was about the same temperature. We will discuss that later with Carl (p. 205). Perhaps the earth was a little bit cooler than it is now, because the luminosity of the sun was less. Assuming a latitudinal variation of mean temperature, I have used 10 to 25°.

Presumably once a large volume of water is lost by dessication, and the remaining puddle or crust is heated up during the days and cooled down during the nights, a high range of temperatures is possible. About 70°C seems reasonable for the maximum, comparable to Death Valley at noon in the sun. A lower bound of about 10° seems reasonable. I assume no glaciation.

There is an entire literature on the pH of soils. Modern soils tend to be acid, sometimes very acid indeed. There are alkaline soils as well. The pH of soils in the absence of bacteria is determined simply by the hydrolysis pH of the silicates that make that soil. They should be about pH 8. The difference between soils and the ocean is simply that soils represent the solid residues left behind. After the water has gone the pH depends very much on the quantity of CO_2 left, as well as the rate at which water goes through the soil. Guessing from soils with very little organic material and very low bacterial counts the pH would be around 6 to 7. This guess is pretty wild too.

I really can't guess whether this is reasonable or not, but I believe the temperature might have been close to the mean annual temperature for the earth.

If a lake or a lagoon which is typically a restricted arm of the ocean, desiccates materials eventually will be blown back into the atmosphere.

A certain amount of bubble bursting occurs in the ocean and in lagoons, which also may involve injection into the atmosphere of salts and gases; there is no reason why organic molecules can not also be injected into the atmosphere by bubbles bursting.

Desiccation, of course, is a very short-lived phenomenon. We must ask organic chemists how long a period of desiccation can be allowed? Can any desiccation, once say, formaldehyde is formed, be allowed before polymerization?

Lakes are relatively temporary phenomena, not long lived. They tend to persist on a time scale of a few thousand years for glacial lakes, to a few tens of millions of years for great lakes like the Eocene Green River Lake. That is probably about the longest time you can expect for a lake to persist which probably gives you plenty of time, presumably when the earth was in some state of tectonic upset.

There may have been a fair number of lakes which would have lasted a long enough time, depending on how long it had to take to accomplish organic synthesis, but sooner or later lakes drain through a river into the ocean. The Eocene was relatively unusual as most tectonic lakes last closer to a million years than to tens of millions of years, as far as I can tell.

Hot springs are interesting because of their extremes. These also are places often associated with acidic volcanic disseminations. The pH values tend to be relatively low, depending on how much HCl, and CO_2, and H_2S comes out. Temperatures may go up to 95°C or even slightly higher.

MILLER: Are there any dry places associated with hot springs where the temperatures go up to 95°?

SIEVER: No, I'm just talking about the water in the little pools next to the hot springs. In a sense, the hot spring cools off when it dries up.

There are thermophilic algae now growing in these hot spring pools, as you know. Once they got started, obviously this was a reasonable place to live. Hot springs also drain into rivers, lakes, and eventually water ends up in the ocean and lagoons.

There is a relationship among the abundance of lagoons, rates of sedimentation, and the tectonic state of the earth's surface. The more the tectonic disturbance the greater the local rates of sedimentation, the higher the likelihood of having lagoons, because of the large sediment supply. Barrier islands, spits and bars form as great amounts of sand, which is in itself a weathering product, accumulates. Those sands form the lagoons.

The lagoons generally have a pH very much like the ocean. They are connected to the ocean, but by a somewhat restricted circulation. Yet because of some increased evaporation in lagoons the pH may be slightly higher. Ultimately, that rise in pH of the lagoons may simply be a biological effect and in the prebiological world lagoons may have had the same pH as the ocean. Yet through desiccation the pH may rise, because of the precipitation of calcium carbonate and calcium sulfate. I have suggested 10 to 25° as rough estimates of the temperature (Fig. 14, p. 186).

It is interesting that lagoons are a trap for clays. Typically at the shore line, sands bar off some lagoons from the ocean and the lagoons operate as still water places where, when the rivers enter, the clays settle out relatively slowly.

The clays will settle through a column of water that may sometimes be up to 100 m deep — Calculations of settling velocities for average clay particles show it may take several days for particles to coagulate and then settle to the bottom in such a lagoon.

This may have been a possible conveyor belt for moving organics out of the ultraviolet transparent zone to the bottom. I want to ask you about adsorption of materials on clays and desorption, which may be a function of both ultraviolet irradiation or any other kind of energy input, plus a pH effect.

I'm wondering if there is a strong pH dependence of adsorption of for example, small organic particles and amino acids on clays, so that the bottom and the top of the ocean may be stratified with respect to pH as well as with respect to temperature. I haven't thought this through much, but we ought to discuss this as a possibility for sequestering organic molecules and putting them together.

ORGEL: I know you don't have the details, but can you give us some idea on the range of humidity in these locations.

SIEVER: Yes. At Death Valley as I recall, the humidity is about a fraction of a per cent relative humidity at the high temperature around 90°, in units of millimeters of water. The authority for that is the occurrence at very low humidity levels of the anhydrous iron oxide hematite, as opposed to the hydrous. The boundaries between those two mineral species is a function of the relative humidity, and there are a number of places in deserts where hematite forms. Another mineral pair that might be used for relative humidity estimates is azurite-malachite, the copper carbonates. Azurite, the less hydrated mineral, is probably a good index of low humidity.

This is a terribly threadbare case, which has been argued among mineralogists for years; there is enough serious question that I can't really say that this is a good way of estimating relative humidity, but it suggests very low values, indeed, in some deserts.

MILLER: Isn't this usually measured with the wet bulb and dry bulb technique in deserts? I have never heard of any measurement as low as half a per cent.

SIEVER: This is a measurement from some place in the Mojave, in Death Valley.

MILLER: A really dry day is 5 per cent and a big thing is made of it. Ordinarily a good dry day is about 30 per cent relative humidity. I just wonder how it would ever get so low.

SIEVER: I offer no authority. This figures comes out of the recesses of my memory. I haven't looked up the data.

BARGHOORN: Death Valley is way below sea level.

MILLER: That is how Los Angeles gets so dry. The Arctic winds come to Los Angeles by way of Nevada, but still there are not one per cent relative humidity days in Los Angeles or San Diego.

PONNAMPERUMA: The extent of your boxes (Fig. 14) for 4 billion years ago, say, can you estimate the total volume of these environments compared to today?

SIEVER: Accepting the model that the oceans were about as large as they are today, it is the major reservoir. Hot springs are transitory and their volumes at any given time must have been very, very small. Lakes may have been moderately extensive, but probably less in total volume than the lagoons.

ORGEL: Is it conceivable that within the same day, or day and night, conditions change from dessication at high temperature to cold moisture at night? Does that happen? In hot deserts apparently there are high temperatures and also dew forms at night, but does dew form on the days on which the temperature has been hottest in the middle of the day?

SIEVER: I don't know.

SHELESNYAK: Yes. Dew forms in the mornings following hot days.

ORGEL: It does?

BARGHOORN: There are weather stations on the coast of Peru and northern Chile where rainfall can not be measured instrumentally, but at night collects on the plants and drips down and supplies enough water to sustain vegetation.

SHELESNYAK: Systems of desert irrigation have been based on this.

ORGEL: In Israel?

SHELESNYAK: Yes. In biblical times pyramids of rocks were built up and at this level enough water condensed for certain types of agriculture, every day, every night.

SCHOPF: Ray, are volcanic areas, lava flows, and so forth, regarded as such transient, small-scale environments that they are not to be considered on your chart?

SIEVER: No. They are essentially fresh water lakes, and fall toward the lower end of the time scale on lakes.

SCHOPF: I mean, lava flows.

SIEVER: The time scale of cooling of a lava flow is of the order of days? So that is the time scale.

SCHOPF: That is a microenvironment considered by some people to have been rather important in certain abiotic syntheses. I notice it isn't on your chart.

SIEVER: I have restricted myself to the abundant aqueous environment. I will include lava as a kind of soil, a soil that stays hot for a day or two.

NAGYVARY: Is it justified to disregard the buffering capacity of sulfides? Hydrogen sulfide might have arisen exactly due to the action of hydrochloric acid or of CO_2. Would you please comment on the possible forms in which sulfur was available and how it might have been recycled three billion years ago? What sort of sulfur distribution do you visualize? Is it possible that it was the interior, and came up at a certain rate?

SIEVER: I assume a fair amount of hydrogen sulfide came out in the gases. It is soluble in water, and immediately would react with any iron that was around. I would presume the stable form would be ferrous sulfide, which usually comes out as a very finely divided black precipitate. At present it is usually in a bacterial formation: hydrogen sulfide is formed by bacteria, but ferrous sulfide is then produced without any trouble. As ferrous sulfide it is an insoluble material, with a very low buffer capacity − lower than, say, the carbonate buffer.

MARGULIS: Might you now project back between 4.7 and 4.5 billion years ago and give us any idea, before this cycle starts, like during the differentiation itself, what kind of differences you would expect? Is that completely impossible?

ORÓ: It's the $64 question.

MARGULIS: Because here we essentially have an abiotic, normal earth.

SIEVER: With the exception of the atmospheric composition.

MARGULIS: Namely, oxygen in the atmosphere. But everything else is the same.

SIEVER: No. There is also a difference in the amount of hydrogen in the atmosphere. If hydrogen is there controlling the gas equilibrium, it does make a difference.

PONNAMPERUMA: How much hydrogen?

SIEVER: Are we ready to start talking about hydrogen escape?

MARGULIS: Do you want to say something about earlier times?

SIEVER: All I can say is that the period earlier than I have already discussed I would call the accretionary period; differentiation must have started as accretion was still continuing. After all, accretion must have been an exponential decay, as we kept sweeping out less and less material into the orbit of the earth. Accretion would continue at a slower and slower rate, and I'm not able to calculate what those rates are, but the accretion simultaneously would build up radioactive heating, which would then allow the sedimentation of the core. Apparently all these processes went on simultaneously.

My guess would be that the surface of the earth was like the moon or Mars at the beginning; a cratered surface with meteoritic infall, and probably the beginnings of volcanoes. Then, sooner or later, the water started falling. The minute water falls, this weathering cycle all goes into operation, since it all really depends on rain.

SANCHEZ: Another surface feature which we are interested in is glaciation and the existence of large masses of ice. What might have been the extent and lifetime of glaciers?

SIEVER: If the temperature were cold enough, and if there were extremes in temperature, then we simply add to this different temperature regimes in these lakes. Glacial lakes and ice pockets only imply changes in the lake temperature.

SANCHEZ: Might these be annual cycles? Are these short cycles you are referring to?

SIEVER: Glacial lakes, if they are relatively large sized in high, mostly northern latitudes, tend to have very narrow temperature excursions. Of course, if they are ponds, they get very warm in the summer. Is that what you mean? The temperatures can go from close to 0° up to about 12 or 14° in a year, and then back down again.

BARGHOORN: Isn't Bob really asking about the fraction of the earth's history that had any glaciation, as opposed to the fraction of the earth's history which was nonglacial?

SIEVER: We really only know about glacial epochs in the last billion years. Is that right?

BARGHOORN: Yes.

SIEVER: And the duration of the glacial epochs, judging from this last one, and what we can deduce from the earlier one, couldn't have lasted much more than a few million years.

BARGHOORN: There could not have been much area that was glaciated.

SIEVER: Nevertheless, glaciers may have provided a really nice, cold place to accomplish synthesis or to temporarily store molecules. The rain falls as snow on glaciers, providing another route of migration. Your materials can be kept in the deep freeze for a while until the ice melts. Of course, the snow falls someplace, and the glacier moves, and eventually it melts at its edges and is then released, in general, to the ocean. That is a perfectly reasonable path to follow, if it is what you need.

The earth, you see, is a great place which can offer a variety of different machines which you can use. What I would like to hear from you is: what is the ideal machine?

LEMMON: Weren't the ancient soils slightly more acidic than the ocean?

SIEVER: That may simply be an artifact of our present day experience in the biological world. If I take some values for soils in a high rainfall period during the year, the same thing is likely to happen that happened with rivers. I give a lower pH for rivers, because waters are loaded with carbon dioxide as carbonic acid, going through so fast that it is sort of like a steady state reaction in which more reactants are constantly being supplied. The pH in equilibrium with that amount of CO_2 is going to be around 5 or 5.5, depending on the precise value. The more water there is, the lower the pH.

LEMMON: I see.

SIEVER: On the other hand, if there is a dry period, and just a small amount of water around, it reacts fully and exhausts the CO_2.

FERRIS: I'd like to get your comments on the question raised about hydrogen in the atmosphere.

SIEVER: The calculations for hydrogen escape were originally based on the temperature of the exosphere, and I believe there is still some question about exactly what the temperature of the exosphere is.

SAGAN: I think it is highly uncertain.

SIEVER: Yes. The whole rate of hydrogen escape depends not only on what the current exosphere temperature is, but what it was likely to have been on the primitive earth.

SAGAN: It depends on the abundance of minor constituents because the exosphere's temperature gets thermostated depending on the conductivity, and the infrared emissivity. A small abundance of polyatomic gases in the upper atmosphere can be a very substantial coolant, but the problem is like an integral equation. If there is a lot of hydrogen, then the high conductivities force the exosphere temperatures to go down.

If there is enough hydrogen, the temperatures tend to be reduced and a lot of hydrogen is retained. If there is little hydrogen, the temperatures are higher which tends to make the hydrogen escape.

Yet I certainly expect the exosphere temperature at the time we are discussing was significantly less than what it is today, apart from any question of solar evolution.

SIEVER: If all this is pushed back to 4.7 billion years ago (a very early date for the differentiation of the earth) and this all happens synchronously: the end of accretion, the blowing away of hydrogen and noble gases, and remembering also that the solar wind was more powerful, much stronger, at that time, I assume, the whole machine was set in motion very early.

Originally the atmosphere was mostly composed of hydrogen, and the question is the rate at which loss occurred. As the dates are pushed further back in time the whole process started earlier, and the start with more hydrogen is made more possible.

The question of CO_2, goes back to the prebiotic synthesis question. If there was too much hydrogen there would have been hydrogen, methane, and no or very little CO_2. The necessity for CO_2 may in fact, be the best way to reason what the hydrogen abundance must have been. Is some CO_2 needed? Is it necessary to have a fair amount of CO_2?

MILLER: Certainly CO_2 is not needed, but on the other hand, it doesn't interfere either.

SIEVER: CO_2 is needed for the weathering cycle (Fig. 14). Weathering on the surface of the earth really does change radically with a methane, hydrogen, ammonia atmosphere.

PONNAMPERUMA: How much CO_2 is compatible with the postulated presence of hydrogen?

MILLER: Will ammonia gas cause weathering?

SIEVER: No, it depends. Ammonia dissolves in water, and gives high pH's and, most of the anhydrous silicates tend to be stable at high pH's, so there would not be destruction by weathering. There would not be much clay formed in the weathering process using ammonia.

MARGULIS: I haven't studied this yet, but in his chapter on the first billion years Holland's estimate of atmospheric CO_2 certainly is up − I think by a factor of ten relative to its present value, and he has detailed arguments for why it was up (Holland, H.D., in manuscript, 1972).

PONNAMPERUMA: But he postulates at the same time one atmosphere methane.

MARGULIS: It is a question of what time, this is not 4.7 billion years ago but 3.5 toward the present. CO_2 was higher and then dropped to present values.

PONNAMPERUMA: During the time when he has now postulated a one meter deep oil sphere (Lasaga and Holland, 1971) he has put an upper limit of ten atmospheres of methane on it — namely during the first half billion years.

BADA: Ray, did you just say that the CO_2 is necessary for the weathering?

SIEVER: Yes, or other acids, such as HCl would work too.

BADA: Maybe HCl was as abundant then as CO_2 is now.

SIEVER: Okay.

ABELSON: HCl is not as nice a weathering agent as CO_2, because CO_2 can continue recycling after the initial move; with HCl that is all there is.

SIEVER: The trouble with strong acids such as HCl is that everything gets tied up with salts, whereas with a weak acid there is still a fair amount of gas around.

MILLER: It is my impression that CO_2 has always been around, because all the high temperature reactions, like volcanoes, tend to shift the equilibria over into CO_2. Electric discharges, and UV give some CO_2. On the basis of plausible differences in pH and calcium ion in the primitive ocean, one might defend a factor of ten higher. However, calcite and aragonite tend to be sluggish in precipitation in the absence of biological systems, and this might raise the CO_2 a great deal above the equilibrium value.

SIEVER: Not terribly sluggish. Calcium carbonate really behaves fairly well abiogenically. It's just slow by comparison in the modern ocean to the way the organisms cause it to behave; but it doesn't get supersaturated by an order of magnitude.

MILLER: Not even by a factor of ten?

SIEVER: It is pretty difficult to take a calcium carbonate solution and keep it supersaturated by tenfold for very long because at just about a factor of ten it starts precipitating giving aragonite instead of calcite. You get the wrong polymorph. Now, sea water is different.

MILLER: Oh! Are you talking about sea water or solution?

SIEVER: I'm talking about an inorganic salt solution. I won't say this about sea water. How supersaturated sea water is depends on how it is analyzed. Organic coatings on the calcite influence the over-all behavior of calcium carbonates.

I wanted to talk about the question of phosphates too but perhaps this should wait until after coffee.

Temperature History of the Earth

ABELSON: Carl Sagan has some arguments about the temperature of the early earth, rising out of considerations of the energy put out by the sun. This perhaps relates to conditions Ray Siever was outlining, and so, perhaps, before discussing further Ray Siever's material we should hear from Carl Sagan. Afterwards Ray can bring our minds back to his discussion of the phosphates.

The Greenhouse Effect

SAGAN: Thank you, Phil.

In an attempt to explain the high surface temperatures on Venus (see, e.g. Sagan, 1960; Pollack, 1969), we have been doing greenhouse calculations for Venus for quite some time. We have also done greenhouse calculations for Mars, but never did any greenhouse calculations for the earth, except to calibrate our calculational techniques.

Just a few months ago it occurred to me that we have all the tools to calculate time-dependent greenhouse effects for the earth (Sagan and Mullen, 1972). Let me say exactly what is meant by "greenhouse effects".

The temperature at the surface of an atmosphere-less planet is an equilibrium between the amount of energy coming in and absorbed and the amount radiated to space: $1/4\,(S/a^2)\,(1-A) = e\sigma T_e^4$. Let us call S the solar constant, so-called because it varies from planet to planet as the distance from sun to planet varies. a is the distance of the sun to the planet in question in astronomical units. S/a^2 is the solar flux in ergs per square centimeter per second at any given planet. We are concerned with the fraction absorbed, so if A is the bolometric albedo, $(S/a^2)\,(1-A)$ is the fraction not reflected back to space. e is the effective emissivity, σ is the Stefan-Boltzmann constant, and T_e is the effective planetary temperature. Since πR^2 intercepts the radiation from the sun, but $4R^2$ radiates to space, there is a factor of one-quarter in the equation.

The bolometric albedo of the earth today integrated over all wavelengths is about 35%. We used to calculate it from measurements of earth light reflected off the moon: "earth shine". But now we know it more directly from space vehicles in orbit around the earth. The solar constant is about 1.94 calories per square centimeter per minute. The emissivity is of the order of 0.9. From this can be derived a surface temperature of the earth. The number comes out about 250°K. This is obviously not the correct answer for the surface temperature of the earth. The explanation of this contradiction is that the greenhouse effect was left out of the calculation. Visible light penetrates through the atmosphere, because the atmosphere is largely transparent. The surface comes to some temperature and emits thermal radiation to space. The left hand side of our equation represents mostly visible light, but the right hand side of the equation is mostly infrared radiation. The atmosphere which was transparent to visible light turns out to be semi-opaque to infrared radiation, and the gases that are particularly opaque in our atmosphere are CO_2 and water. An impedence to the escape of infrared radiation to space results and the surface temperature must increase, so that the fraction of radiation which escapes through atmospheric windows in the infrared equals the amount of sunlight that comes in.

Knowing the amounts of carbon dioxide and water in the earth's atmosphere, the infrared absorption spectra and the fraction of radiation absorbed, one can calculate that the greenhouse effect amounts to some 35°. This is indeed the difference between the observed temperature and the temperature calculated without considering the greenhouse effect. This increase in temperature is correct: the mean surface temperature is about 288° Kelvin. Since this is the correct global average temperature for the earth, everything is fine.

ABELSON: When this greenhouse effect is operating, supposedly a water molecule absorbs some energy, and of course, part of it is re-emitted back to earth.

SAGAN: Let me discuss that. For this purpose the simplest calculation is good enough. You have some spectra (Fig. 15) showing transmission versus wave length for the Earth's atmosphere, with detailed band structure. For purposes of convenience let us consider a step function absorption. With some spectrum such as that in Fig. 15, in certain regions there is a lot of absorption and very little or none in other regions, called windows. The calculation is done by taking the emission to space in a window as from the surface. The emission to space in a region of absorption is taken as from the tropopause. Since the atmosphere tends to be isothermal above the tropopause, a mistake about the exact level will not be too big a mistake. With such a crude square wave model, we can get the correct answer.

Let me explain the relation of the greenhouse effect to early earth temperatures. First I should say a few words about models of the evolution of the sun. The theory of stellar evolution is in quite good shape. If the effective temperatures of all the stars in the sky are plotted versus luminosity you find instead of a scatter diagram a tendency of the values to cluster in a few parts of the diagram, particularly in a thinnish stretch called the main sequence. This intuitively unlikely result is now extremely well understood. Stars tend to live on the main sequence of this Hertzsprung-Russell diagram as they are stably converting hydrogen to helium by thermonuclear processes. Another little cluster is called the red giant region and another is the white dwarf region, each is understood in terms of the nucleosynthetic processes.

This same theory which explains remarkably well the Hertzsprung-Russell diagram, implies that the sun has been gradually increasing its

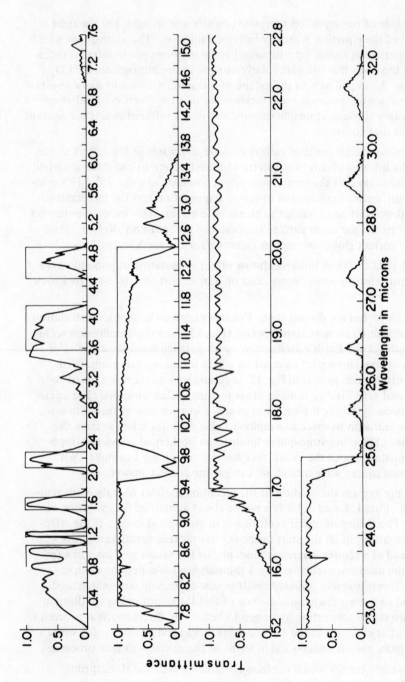

Figure 15. Adopted square wave approximation, derived from laboratory CO_2/H_2O absorption spectra, to the observed mean transmittance of the terrestrial atmosphere. The 9.6 μ feature, due to O_3, was ignored. The correct mean surface temperature of the Earth is derived for a variety of reasonable square wave fits.

luminosity over the last 5×10^9 years. Stellar evolutionists consider this to be a quite reliable result — the sun has increased about half a bolometric magnitude which is several tens of per cent in absolute luminosity over the past 5×10^9 years.

There will always be some uncertainties about this, for example, related to the possibility that the fundamental constants of physics are changing with time. Dicke (1962) at Princeton proposes the Newtonian gravitational constant has changed with time. The stellar luminosity depends on some highish power of G. If that is true then there is considerable uncertainty in what I am about to say.

SIEVER: In which direction would Dicke's work change your results?

SAGAN: To make the early solar luminosity higher than what I suggest. If you were to dislike strongly my argument you could believe that G is a secular variable. Most cosmologists think G is a true constant.

It is also possible that the failure to detect the predicted ^{8}B neutrino flux from the sun may force revision of some concepts of solar evolution, but most stellar evolutionists believe this to leave the solar luminosity evolution unaffected.

Anyway what I will tell you has some probable error attached to it, but I think the solar evolution assumed has a high probability of being correct.

LEMMON: Carl, might you briefly explain why the luminosity over the last five billion years is slightly increasing, rather than slightly decreasing?

SAGAN: The trend is toward the red giant stage where there is a great increase in luminosity. The increase in mean molecular weight of the sun by the conversion of H to He is an important factor. I'm sorry not to be able to give you a thorough explanation in just a few words!

LEMMON: But it is good enough. Thank you.

SAGAN: There has been variation, both in the solar luminosity and the radius with time. Okay? As you can see, the solar luminosity increases by tens of per cent, but the radius is also increasing.

Using our simple radiation balance we have calculated first the airless earth's temperature as a function of time with that luminosity curve, and then the various greenhouse effects. Obviously, our first interest is today's kind of CO_2-H_2O greenhouse effect.

On the next slide the ordinate shows time in aeons — billions of years. The abscissa gives temperatures (Fig. 16).

*Figure 16. Calculated time-dependent model greenhouses for Russell-Bond albedo $\overline{A} \simeq 0.35$
and two surface infrared emissivities, $e = 0.9$ and $e = 1.0$, the former being more nearly
valid and giving the correct present global temperatures. The CO_2/H_2O atmosphere
assumes present abundances of these gases, pressure broadened by 1 bar of a foreign gas.
The slightly reducing atmosphere has the same constituents with the addition of a 10^{-5}
volume mixing ratio of NH_3, CH_4, and H_2S. In this case ammonia is the dominant absorber. At the top is shown the greenhouse resulting from the addition of 1 bar H_2 to
the constituents already mentioned. The evidence for liquid water at 2.7 to 4.0 AE ago
comes from a variety of geological and paleontological data. The time-evolution
of the effective temperature is also displayed. All calculations are for $\Delta L = 30\%$; for
larger time-derivatives of the solar luminosity, the freezing point of seawater is reached
in yet more recent times assuming present atmospheric composition.*

The lowest curve is the effective temperature determined by this equation. The present value is about 250°K. As you go back in time that temperature declines in a major way, so that 4.5 billion years ago, if the earth had no
atmosphere, the surface temperature would have been about 230°K.

This calculation (Fig. 16) shows the CO_2-H_2O greenhouse effect with precisely the proportions of CO_2 and water in today's atmosphere. I want to discuss the possibility that those proportions were different.

You can see that the present surface temperature of the earth is about
285°K which is the correct value.

As we proceed back in time, assuming the atmospheric composition has
remained identical, we find we cross the freezing point of water at about 2
billion years ago (Fig. 16). These are global average temperatures. At 3 or 3.5
billion years ago, the temperature was roughly 10°K below the freezing point
of sea water.

Since at both 1.5 and 3 billion years ago there is biological, paleontological and geological evidence for liquid water, we have a serious discrepancy.
And as we go back toward the time when some of us like to believe life was
originating we find a mean surface temperature of the earth was 10-15° below the freezing point of sea water.

Since these are global average temperatures and there is significant latitudinal temperature variation, there may have been a thin equatorial region
where water was just barely above the freezing point. In general the planet
must have been frozen.

There is good evidence for rather warmish conditions during the Precambrian. The planet was not on the verge of being frozen all over. It is unlikely, e.g., that *all* samples of Precambrian algae are from some thin equatorial latitude strip. Therefore, I believe that a global average temperature below
the freezing point of water contradicts the evidence.

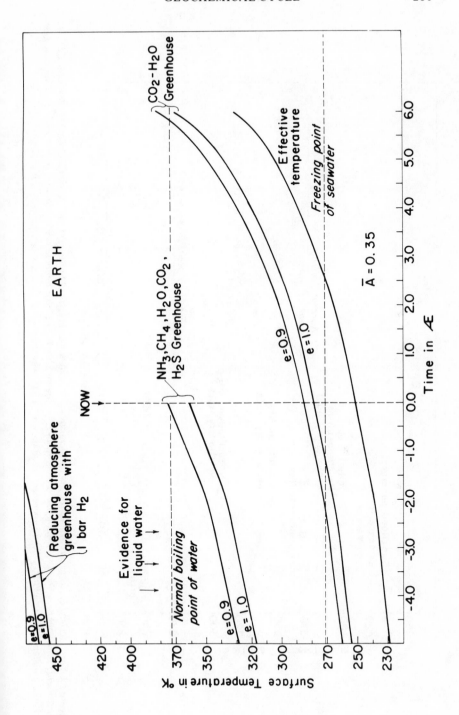

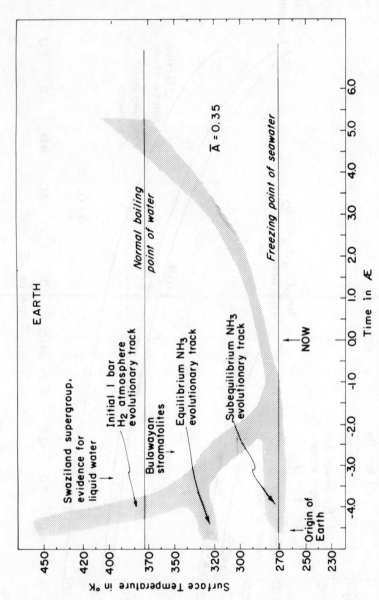

Figure 17. Three derived evolutionary tracks for the temperature of the Earth. As described in the text, the subequilibrium NH_3 track ($< 10^{-5}$ volume mixing ratio), while schematic, is thought to be most likely. The amplitude of global temperature oscillations in Mesozoic and Paleozoic glaciations is smaller than the width of the temperature curve shown. A runaway greenhouse occurs several eons in our future.

MILLER: What's the temperature spread today between the equator and the poles, or between the equator and the average?

HULETT: I think it is about $50°$ centigrade.

BARGHOORN: The equatorial average is about $80°$ and the border of ice sheet is about $32°F$.

SAGAN: It's confusing enough without Fahrenheit! The global temperature is something like 285 K, which is about $12°C$. The equatorial average is about $15°K$ higher.

Now, I have assumed that the albedo of the earth is independent of temperature, but, obviously, this is not true. There is an instability well known to people who worry about ice ages: as the temperature is reduced, the fraction of the earth which has polar ice caps is increased. Therefore, the global albedo rises, and the temperatures lowers even more. This is one reason why I think these are probably upper limits for the temperature.

As the mean temperature goes down, obviously the vapor pressure of water lowers and the amount of water vapor in the atmosphere is less. Therefore the greenhouse effect due to water vapor in the atmosphere will be less than I calculated. A similar remark applies to CO_2. Therefore I think the global temperatures could have been lower than I calculate and the actual discrepancy may be significantly greater. The present average albedo is 35 per cent, but freshly fallen snow has an albedo of 80 per cent. Higher albedos lead to quite significant lowerings of temperatures.

What can we do to rectify this contradictory situation? Let's go by small steps, and make small perturbations to the existing situation. Can we increase the temperature via the greenhouse effect by adding more CO_2 to the atmosphere?

The trouble is that even though CO_2 is only 0.03 per cent of the atmosphere, CO_2 absorption tends to be already saturated in the regions where it is strongly absorbing. Doubling the amount of CO_2 does not change the absorption very much. This is now well-known to people who worry about causing a runaway climatological effect on the earth by burning fossil fuels.

SIEVER: Is this generally agreed on?

SAGAN: Yes. Increasing the CO_2 abundance by factors of 2 or even more will not make any significant change on the greenhouse effect.

ABELSON: How about lowering CO_2?

SAGAN: Eventually if the amount is lowered the CO_2 bands will be desaturated and then the temperature will lower, but pushing it down slightly will not change anything because we are not sitting right at the precipitous edge. But if CO_2 is reduced this discrepancy of the low calculated temperature is enhanced even further. Also CO_2 is very strongly buffered. It's hard to change its abundance.

SCHOPF: What happens if CO_2 is increased an order of magnitude more? There have been suggestions that perhaps CO_2 was that much more plentiful.

SAGAN: Yes, an order of magnitude more CO_2 will help, but will not remove the bulk of this discrepancy. It will help but not nearly as significantly as some other molecules I'm about to mention.

SIEVER: Are you saying the amount of increase in greenhouse effect of a very few parts per million of industrial CO_2 that people are worried about is negligible, but if CO_2 was increased by an order of magnitude this might cause a significant change?

SAGAN: Even a largish increase in the amount of CO_2 would not resolve our discrepancy. Venus is an excellent case. On Venus there is 100 atmospheres of CO_2, but the bulk of the greenhouse effect is not due to CO_2.

Obviously, I can fill in any spectrum if I have infinite latitude in pressure, but my point is the insensitivity of the global temperatures to a CO_2 abundance anywhere around the contemporary value. The CO_2 abundance is highly controlled, by silicate-carbonate equilibria; by buffering with seawater which contains almost 100 times the atmospheric CO_2, and by the respiration/photosynthesis feedback loop.

Ammonia: Possible IR Absorber in the Early Atmosphere

We must look at other reasonable constituents. Oxygen and nitrogen, being homonuclear/diatomic molecules, have no permitted rotation-vibration structure. Noble gases obviously won't help. Some gases like CO, which might have been around, don't happen to have significant absorption features. They are either too narrow, or in regions which are already absorbed by CO_2 and water. H_2S doesn't help very much. Neither does SO_2, ozone, nor oxides of nitrogen.

Obviously somewhere here we must worry about methane and ammonia; Methane and especially ammonia are splendid molecules for greenhouse effects in terrestrial planets. Ammonia is lovely because it has an exceptionally strong absorption band centered at ten microns, which is the peak of the thermal emission to space from the earth at its present temperature. The Wien maximum of the Planck body distribution from the earth is centered on ten microns. Obviously the plugging of that window, where most infrared energy is pouring out, will significantly augment the greenhouse effect. Since the strongest ammonia band is at 10 microns, a large quantity of ammonia is not required.

We have calculated the greenhouse effect for atmospheres with present abundances of water and carbon dioxide, a few parts per million of H_2S and CH_4 (which doesn't change anything at all) plus ten parts per million of ammonia. We used ten parts per million of ammonia because it is consistent with the Bada-Miller argument, and secondly, it corresponds to about ten centimeter atmospheres, a convenient quantity for laboratory infrared measurement. Thirdly, as you will see, 10 ppm is enough to do what we want.

I stress that for our purpose, i.e., raising the temperature of the earth above the freezing point of water, only minor concentrations of ammonia are needed. Furthermore, these concentrations are consistent with two independent arguments of Bada and Miller (1968).

Their ammonia abundance is set independently by its solubility and mineral interaction in water, and by the amount required for the prebiotic synthesis of asparate. These seem now to provide rather interesting limits on the ammonia concentrations in the atmosphere.

As for methane, I don't know what sort of limits there are on it, Cyril suggested one atmosphere of methane.

PONNAMPERUMA: Holland's figures for the first half a billion years were from one to ten atmospheres. Although done five years ago, I think he stands by them even in the latest chapter of his new book (Holland, 1972).

SAGAN: I just looked through that, and all I found was his statement that there was more methane than CO_2. But if what you and he both say is right, he is talking about a great deal of CO_2 too.

MILLER: I have never seen any figure given for the amount of methane and, as you know, I feel that it ought to be high. But some people, even in this room, believe methane is really not a suitable molecule at all for prebiotic syntheses.

SAGAN: In any case, methane happens to be a molecule rather like CO_2. That is, significant changes in the abundance will still not make any major changes in the absorption spectrum. I have again put in a few, or ten parts per million of methane.

ABELSON: What happens if you add a few parts per million of N_2 or NO_2?

SAGAN: N_2 does nothing whatever. Oxides of nitrogen are ineffective. Okay. We have done the calculation for ammonia, methane, water, CO_2, and H_2S as constituents and in a moment I'll discuss the relative contributions of each (Fig. 17). Obviously, with the present sun and these gases you can get very high temperatures, but that is not the issue.

HULETT: Is that calculated on the present amount of water?

SAGAN: Yes. Smaller water abundances require more ammonia. As you will see, ammonia is the main story. In Fig. 17 we see 3 evolutionary tracks. I here neglect the one with 1 bar initial H_2. The middle track has the Bada-Miller equilibrium amounts, $\sim 10^{-5}$ NH_3 mixing ratio. The lowest track has $\sim 10^{-6}$ NH_3, a value more consonant with the photodestruction of NH_3 by ultraviolet light.

This morning we learned why low temperatures are desirable at the time of the origin of life. The prevention of degradation of various products, the increase of several orders of magnitude in the concentration of organics in the ocean is a very major factor. I find it pleasing to come out with temperatures just above the freezing point of water shortly after the formation

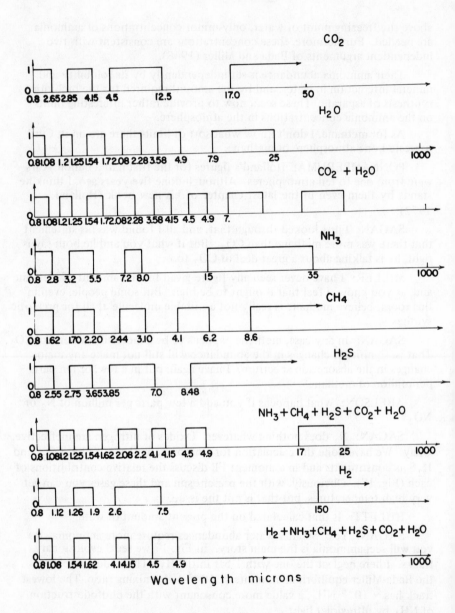

Figure 18. Square wave approximation to the absorption spectra of some of the candidate atmospheric constituents investigated, based on laboratory spectra (courtesy of Dr. B. N. Khare) and theoretical calculations (courtesy of Dr. L. D. G. Young). CO_2 and H_2O are represented for their present abundances, CH_4, H_2S and NH_3 for volume mixing ratios $\sim 10^{-5}$, all for 1 bar foreign gas broadening. The H_2 spectra are for 1 bar of H_2, self-broadened. Various combination spectra are also exhibited.

of the earth, and temperatures which are mildly tropical about 3 or 3.5 billion years ago.

MILLER: But that implies that the Buluwayan limestone —

SAGAN: It does imply methane and ammonia at that time. These two curves (Fig. 17) must join in some way that doesn't cross the freezing point of water. They join as the reduced compounds disappear due to the escape of hydrogen, and photodissociation such as Ray Siever mentioned. Also these results are consistent with some small fraction of ammonia in the atmosphere in *Kakabekia umbellata* times. Siegel (Siegel & Guimarro, 1966; Siegel *et al.*, 1967) finds this microbe which looks like a facultative ammonia user, 2×10^9 years ago. I have found this to be an entertaining coincidence as well.

PONNAMPERUMA: He found Kakebekia on the top of a Hawaiian volcano recently.

SAGAN: The same organism? It doesn't live only in urinals?

YOUNG: He's found it on islands.

SAGAN: Has it a global distribution?

YOUNG: It is not an obligate ammonia user either.

BARGHOORN: Facultative. It's tolerant.

SAGAN: It holds it's breath.

YOUNG: You would too.

SAGAN: This shows (Fig. 18) what happens just for methane, ammonia, and H_2S. The CO_2 greenhouse added to the methane-ammonia-H_2S greenhouse, gives more than the total, because some bands fall on other bands; it is not a linear superposition process.

You see, if the molecules absorb where there is no infrared radiation anyway, nothing is gained.

NH_3 is much more important than any of the others. H_2S is the least important. If all the H_2S disappeared nothing I have said will be changed.

MILLER: Carl, many of us always speak about methane, but others have suggested that perhaps other hydrocarbons might have been even more abundant; for example, acetylene, which is produced by electric discharge or ethylene.

SAGAN: I haven't looked into the absorption properties of hydrocarbons, but I do not think their effect will be significant.

ORGEL: If all the windows were filled with absorbing gases how hot would the surface of the earth get?

SAGAN: As hot as the surface temperature of the sun.

SCHOPF: Would higher molecular weight hydrocarbons be expected to cause this greenhouse effect in general?

SAGAN: A primary absorption feature of all hydrocarbons is the CH stretch at 3.5 microns – but methane is already just bombing that out. I suspect that any plausible hydrocarbon will have a methanish absorption.

ORGEL: I think you might begin to fill up all the gaps, the whole damn lot, with an appropriate mixture of hydrocarbons.

SAGAN: No. The major window in the gas mixture I have discussed is 8 to 14 microns. To increase surface temperatures even by a small amount, say 10°, you need something which plugs the 8-14 micron hole.

ORGEL: What are the wave numbers?

SAGAN: A thousand wave numbers is 10 microns.

ORGEL: Can't that window be plugged with a modestly long hydrocarbon?

SAGAN: Pure rotation spectra are generally needed.

ORGEL: No. A pure rotation spectrum is way off, this is 300 wave numbers.

SAGAN: No. The pure rotation spectrum is exactly what fills in there.

ORGEL: Where is the gap in frequency compared with the center of your ammonia absorption?

SAGAN: About twice the wave length, half the wave number.

ORGEL: Is it a pure rotation band of ammonia that is absorbing?

SAGAN: No, at 10 microns it is a vibrational band.

ORGEL: Twice a vibration of ammonia is still a vibration.

SAGAN: No, the first overtone of a 10 μ fundamental absorbs at 5 μ. Overtones don't help, they go the wrong way. Leslie, what molecule do you think can do it?

ORGEL: I figure you are talking about between 300 and 500 cm^{-1}, which is in the middle of the infrared. There are lists of organic molecules which absorb in that region, and their absorption has nothing to do with rotation, which is about a hundred times less intense.

SAGAN: Leslie, you are wrong!

For example, at 20 to 30 microns, the pure rotation spectrum of water exists, but the lines are very narrow and separated. The pure rotation spectrum of ammonia begins at about 30 microns, and that is what starts cutting off radiation at long wavelengths in this spectrum. This can even be calculated from first principles –

MILLER: Carl, I am not very familiar with infrared absorption spectra, but I think it is reasonable to consider substantial quantities of other hydrocarbons. You might wish to include alternatives.

SAGAN: Fine. Let me stress again that serious perturbations on these calculations imply absorption in the 8 to 14 micron region, a highly non-trivial problem.

LEMMON: How about HCN?

SAGAN: It doesn't help. There are some planar deformations of ring compounds. But they are several orders of magnitude less intense than the fundamentals were are discussing. They are fundamentals, but a planar deformation fundamental is several orders of magnitude weaker than a vibrational fundamental.

ORGEL: No, maybe down by a factor of two, but not orders of magnitude.

SAGAN: Leslie, not at all.

ORGEL: You can look at a sheet of paper on an infrared spectrometer on which the absorption is measured, and it runs from about 30 mμ down to 5 mμ.

SAGAN: Leslie, I'm afraid the problem is that you are not doing reciprocals correctly!

ORGEL: That may be correct.

BUHL: Carl, in the present earth's atmosphere isn't it blanked out between 20 and 30 microns?

SAGAN: There are some windows, a 20 micron window, for example –

BUHL: And an 8-13?

SAGAN: Yes, traditionally used to look at planets and so forth.

ORGEL: I have probably got a factor of ten out of my reciprocals.

SAGAN: Very likely. There must be some simple explanation!

Okay. Here is the square wave approximation to the absorption in the atmosphere for CH_4 plus ammonia plus H_2S plus CO_2 and H_2O. We have no absorption, complete absorption, no absorption, complete absorption and so on.

Do you want this in immense detail?

ORGEL: I want to know what you think 18 microns to 25 microns are in reciprocal centimeters!

SAGAN: OK: 10 microns is 1000 wave numbers. 20 microns is 500 wave numbers.

ORGEL: Yes, and do you believe that there is a pure rotation spectrum in that region?

SAGAN: Yes, I do.

ORGEL: It's new to me.

SAGAN: Evidently!

ORGEL: But I assure you there isn't!

ORÓ: Leslie, carbonatites are new to him!

SAGAN: Here is the composite spectrum from 0.8 to 4.9 microns, here is 1 micron, which is 10,000 wave numbers.

ORGEL: I'm beginning to see that you are talking about the upper overtones. You are quite right for the overtones, but not for the fundamentals.

SAGAN: Let us agree on some fundamentals: 3.33 microns, where there is complete absorption by methane. There is complete absorption from 4.9 to 17 microns, and this is the major contribution to this greenhouse effect. The longest wavelength contribution is the ν_3 fundamental at 15 microns of CO_2, which is the longest wavelength vibrational fundamental in our set of gases.

ORGEL: Right.

SAGAN: Starting at about 13 microns is the ammonia vibrational fundamental, and then methane and ammonia absorption. 17 to 25 μ is transparent. At 25 microns the pure rotational spectrum of water comes in, which is happening in our present atmosphere, and at slightly longer wavelengths the pure rotation spectrum of ammonia.

ORGEL: I think you will also find that there are intense modes, due to bending of carbon molecules containing three and more carbon atoms in here.

SAGAN: That is very different from ammonia.

ORGEL: Right.

SAGAN: I would guess that the fraction of atmospheric molecules with three or four carbons would be less than the fraction with methane. The individual transition probabilities are also down so I think it unlikely that larger molecules would fill in this region.

If only a little bit of the 17-25 μ region is filled not much will be changed, but if a significant fraction can be filled then radiation escape will only be at 4.9 microns, far down –

MILLER: What molecule could absorb in that window, Carl?

SAGAN: Hydrogen at a pressure of about an atmosphere, an atmosphere's worth of hydrogen.

PONNAMPERUMA: Carl, will you calculate the Rasool atmosphere and the rate of escape with this sort of molecule?

SAGAN: Such atmospheres have hydrogen exospheres. The thermal conductivity of hydrogen is so great that they are very cool and remain for long periods – about 10^9 years.

It is extremely difficult to make any sizable perturbations in these temperatures under my assumptions. Obviously refinements may be done, but our experience shows this will not change the results by a significant amount. The theory of stellar evolution that forces this set of conclusions seems to me moderately plausible. It implies that the global average temperature of the earth (from exogenous radiation only) at the time that the earth formed was around the freezing point. The temperature then increased to several tens of degrees centigrade as the sun evolved up the main sequence, and then there was a transition from that reducing atmosphere, with trace constituents of ammonia —

LEMMON: How was the temperature affected by endogenous heat sources?

SAGAN: I will discuss that next.

BADA: If your error in solar luminosity was 10 per cent what would the contribution of just CO_2 to the greenhouse effect be? From your first diagram the luminosity varied over the last five billion years. How much did it vary, 30 or 40 per cent?

SAGAN: Yes. For a pure CO_2 greenhouse with just the present CO_2 abundance. At the time of the origin of life the surface temperature would have been $250°K$.

BADA: Yes, you are $10°$ too low. If you have 10 per cent error on your luminosity, I calculate you could get your temperature back down.

SAGAN: The errors in the model calculation tend to go the other way. The range of published values for luminosity are between 30 and 50 per cent and I have taken the 30 per cent value.

BADA: Can't there actually be an uncertainty extrapolating from the Hertzsprung-Russell diagram?

SAGAN: No. It doesn't scale that way. There are compositional differences between stars which make some difference in the H-R diagram, but that does not affect the evolution of the sun.

FERRIS: How much effect would changing methane and ammonia by factors of ten make?

SAGAN: Raising ammonia by a factor of 10 from 10^{-5} to 10^{-4} would be significant. But I have believed in the Bada-Miller discussion.

FERRIS: If you change CO_2 by a factor of ten, it doesn't cause much change?

SAGAN: It's a fair change but CO_2 is buffered.

MILLER: One shouldn't take this ammonia calculation that seriously. A reasonable range of partial pressure around these values ought to be calculated.

BARGHOORN: There are a lot of different lines of evidence sea ice, snow line depression in the tropics, the depression of certain tropical forest plants, of large numbers of species in Borneo, and work not yet published by a student of mine in Panama, that the earth's temperature at the equator was about 5°C lower at the time of maximum glaciation than it is today. Five degrees is a high percentage of the total temperature change since the early Precambrian. This is squeezed into one or two million years.

SAGAN: Yes. There are no ice age perturbations on these calculations but they are less than the thickness of the stippled areas in Fig. 17.

BARGHOORN: This is a fact, inferred, but from a lot of evidence from various sources.

SAGAN: Yes, but the temperature goes back up again, doesn't it?

BARGHOORN: Right now we don't know whether we are in interglaciation or not. From the present time, extrapolating back to the time of maximum glaciation, there was an approximate 5°C reduction in the temperature at the equator. The inference for the middle latitudes is large, about a 15°C decrease. Do you think there is any chance that the sun is a variable star?

SAGAN: This is one theory of ice ages which Ernst Opik in particular has developed, but most people in the field don't agree.

BARGHOORN: If it is a variable star, can you with impunity put the sun on the main sequence, just as a steady evolution?

SAGAN: I don't think that is a very serious objection. The long-term expectation for the evolution of the sun's temperature is one thing, and the perturbations on it another. But if we had absolute knowledge that the ice ages were due to spikes in the solar luminosity, then the stellar evolutionists would have to go back and ask: how serious a change is that in our over-all model of the evolution of the sun?

There are many good aspects of present solar evolutionary models, like the age of the sun: 5×10^9 years. That age is totally independent of any observations of ages of earth or meteorites or moon rocks. The theory fits very nicely with other astronomical and terrestrial observations.

What if a change of 10 per cent was made in the delta luminosity? Nobody says that 30 per cent is right, and 40 per cent can't be right. There is not that kind of accuracy. Several tens of per cent is the order of magnitude. So I have taken some representative values. Bill Schopf has been saying I should put error bars on these values.

SCHOPF: Even if you do put error bars around each of those points the qualitative conclusion stays essentially the same. It makes a difference in freezing point. It may be off by as much as 500,000,000 years on each plot, the qualitative argument still stands.

SAGAN: That is precisely the point. Note that Fig. 17 also gives some idea of the epoch of the transition from a reducing to an oxidizing atmosphere.

BARGHOORN: There should be fossil frost features in Swaziland.

SAGAN: Yes, and the absence of fossil frost features shows that there was a greenhouse effect greater than the present greenhouse effect. That's my argument.

If there were only a CO_2 and water greenhouse effect then at the time the Swaziland sediments were formed you would have had to have evidence of frost (Fig. 17). The lack of evidence for frost and ice, and the quite definite evidence for liquid water means that the temperature was higher than can be accounted by today's greenhouse effect and therefore provides evidence for ammonia and methane in the atmosphere.

The transition period from the reducing to the oxidizing atmosphere seems to have occurred between 3.0 and 1.5 billion years ago. As I understand it, my estimate is consistent with the geological evidence.

SIEVER: The temperature 4.5 billion years ago was sufficiently close to the freezing point of water so that any latitudinal variation implies the presence of polar ice caps.

SAGAN: Sure! There certainly would have been polar ice caps.

HULETT: When the sun's temperature was down a bit, how was the UV flux affected in the relevant region?

SAGAN: I have calculated this. Assuming a black body spectrum the flux is lower by no more than a factor of 2 or 3. At short wavelengths, less than 2000 Angstroms, this estimate depends on the details of the chemistry at that temperature in the sun. I promised to discuss the influence of hydrogen. One atmosphere of H_2 is strongly absorbing due to pressure-induced dipole and permitted quadruple transitions. It is absorbing right through this region. An atmosphere of hydrogen just perfectly plugs the big 17-25 micron window. The conclusion is that if there ever was as much as an atmosphere of hydrogen then even higher temperatures are possible.
The absorption is strongly dependent on the total pressure; if it were down to a tenth of an atmosphere of hydrogen it wouldn't matter.

SIEVER: That implies that at some early stage there was a relatively high temperature and a high greenhouse effect from the hydrogen. And when hydrogen was lost cooling set in.

SAGAN: It depends on the scenario. The relative fractionation of non-radiogenic noble gases means that very high mass numbers escaped very efficiently at some early time. That time could not have been after the earth formed, because an exosphere temperature of hundreds of thousands of degrees is needed, which doesn't seem plausible. Therefore, events occurred during the formation of the earth, and that is when the bulk of the hydrogen was lost.

Since the time of the formation of the crust, it is possible that there never was a great excess of hydrogen. It is also possible, as far as I know, that there was one atmosphere pressure of hydrogen that escaped to space during geological time.

SIEVER: It really depends on how much hydrogen was occluded.

MILLER: Do you really think there was a partial pressure of hydrogen of one atmosphere for any length of time? Isn't the rate of escape proportional to the partial pressure?

SAGAN: The rate of escape is proportional to the hydrogen number density at the base of the exosphere, but the time scale for escape is independent of the number density. If there was one atmosphere of hydrogen at the origin of the earth, it could have been lost in a couple of billion years. If there was much hydrogen, it would decrease the exosphere temperature, so hydrogen tends to keep itself. There is a possible scenario with a lot of hydrogen and high temperatures. If five orders of magnitude in xenon is depleted then at least as many orders of magnitude are depleted in hydrogen. Therefore there were quite low partial pressures of hydrogen, determined by some photochemical or geochemical equilibrium with other compounds. I feel that is the more likely situation.

LEMMON: Would you please say something about the endogenous sources of heat on the earth and how they have changed?

SAGAN: I believe the earth was fully melted at the time of its origin. This implies melting temperatures, at the very beginning, of 1200 to 2000°C.

An old calculation going back to Kelvin's time shows that a convective earth at melting temperatures radiating to space can cool in a fantastically short period, like 10,000 years. The reason is just T^4; the radiation is extremely efficient.

LEMMON: But what about the radioactivity?

SAGAN: Whatever that initial high temperature was due to, the earth melted. In the process of melting and differentiating, the large ionic radii, radionuclides like potassium and thorium, concentrated in the surface of the earth. K^{40} is reduced in abundance by several half-lives. The amount of heating is up by a factor of 10 or so, due to the difference in radioactive decay between then and now. Local remelting over the earth for the first billion years of history seem perfectly reasonable, although not absolutely compelling. It gives us an explanation of the lack of any rocks older than 3.5 billion years; namely, every spot on the earth was melted once more during that first billion years of earth history, even though the whole earth was not melted in that period.

LEMMON: How much of a rise in temperature is caused by radio active material in the crust?

SAGAN: Negligible.

The amount of energy coming up the geothermal gradient today is roughly 2×10^{-5} of the amount of energy coming from the sun. It only contributes to, say, heating on the night side of the planet Mercury. It makes no contribution to the surface temperature here.

SCHOPF: One last question. Donn, Donn, and Valentine (1965) made a similar argument, published in GSA Bulletin. As I recall, they came up with somewhat different conclusions. Would you please comment on that?

SAGAN: They even said that surface temperatures were lower and asked if a greenhouse effect could raise them. But they unaccountably concluded that the greenhouse effect can't do it.

ORGEL: Did they think of ammonia as a possible absorber?

SAGAN: They may not have mentioned it explicitly. But once stated, they rejected it with no calculations. Then they concluded that something is wrong with the theory of star evolution. So I wasn't very impressed by the paper.

BARGHOORN: I asked Francis Birch about this not long ago, and he gave me a figure of 1.88×10^{-6} calories per square centimeter per second. And the solar constant is 1.9, what?

SAGAN: The solar constant is about 10^6 ergs per square centimeter per second. Do you want calories per square centimeter?

BARGHOORN: No, the relationship here shows the endogenous heat flow as being absolutely negligible.

SAGAN: Certainly. It is trivially negligible.

FORM OF PHOSPHATE ON THE EARLY EARTH

SIEVER: I have got one minute left for the phosphate story and that is what it will take.

Assume there was a lot of bicarbonate in the ocean, slightly higher than today. That is why I put pCO_2 at some time suitable for us at a slightly higher level. There may have been slightly higher contributions. I simply ascribe to this, possibly, a slightly higher rate of outgassing or recycling at that time. Calcium phosphate is a slow to form inorganically, it's much more recalcitrant than calcium carbonate. Calcium carbonate is not bad in its approximation to equilibrium.

If bicarbonate were higher in the ocean what were the most insoluble fast-precipitating compounds? They turn out to be calcium carbonate, ferrous sulfide, and ferrous carbonate, if the redox potential is kept low. All the ferrous compounds seek their insolubles and (as we are now beginning to learn) ferrous silicates, which also have low pK's, precipitate out moderately rapidly, although they apparently go to some supersaturation.

It is also possible that a ferrous phosphate, not precisely equivalent to, but probably something like mineral vivianite, $Fe_3 PO_4 \cdot 2H_2 O$ can form. In any case there probably would have been slightly higher levels of phosphate dissolved. When there are slightly higher levels of phosphate, phosphate starts adsorbing on available silicate clays. I have found appreciable phosphate adsorption by clays in experiments in my own lab (unpublished) and there's a rather extensive report on the adsorption of phosphate on clay mineral surfaces. There seems to be a fairly nicely reversible adsorption isotherm. It seems to come to near equilibrium within 24 hours. It adsorbs, desorbs, resorbs. These are at pH values from roughly 5.5 to 8.

If there were a slightly higher level of pCO_2, which led to a slightly higher level of bicarbonate in the ocean, then a certain greater amount of phosphate adsorbed on silica in these structures was entirely possible.

The interesting part to me is that this occurs at the pH of sea water. desorbs relatively easily, so that it doesn't get terribly bound up in insoluble silicate structures and become unavailable. I'm just suggesting this as a possible way out of the phosphate problem.

MILLER: You have got three times its saturation, 3×10^{-6} molar phosphate in equilibrium.

SIEVER: I would say an order of magnitude is allowable.

MILLER: So that is 3×10^{-5}, which is really a pretty low value for reactions in solution.

SIEVER: I would have to try to make some fly-by-night calculations about what it might be.

ORGEL: Is carbonate apatite an important phase?

SIEVER: Yes. A good many precipitates made in the laboratory are carbonate apatite. The question of exactly what that mineral is, is not solved, but that needn't concern us. It is something that is being precipitated from mixed bicarbonate-phosphate solution and might be expected to coprecipitate with a bit of calcium carbonate.

ORGEL: The carbonate apatite is a good catalyst for one of the reactions.

MILLER: It is very hard to make apatite without containing some carbonate.

SIEVER: It depends on the rates of the precipitations in a nonbiological world, which I'm not sure of.

MILLER: Phosphate (apatite) binds calcium very strongly. Compounds bind calcium more strongly. One possibility is oxalate, which is not considered a prebiotic compound.

FERRIS: A lot of it can be gotten from hydrogen cyanide.

ABELSON: Not oxalate.

MILLER: Presumably it could come from cyanogen at the proper pH conditions.

FERRIS: Polarizing hydrogen cyanide solutions gives oxalic acid.

MILLER: Another possibility is that imino-diacetic acid which was synthesized in the electric discharge experiments –

SIEVER: Yes, but that has a pK of about two orders of magnitude.

MILLER: But the thought is that there is a concentration of the chelates; I mean, say apatite is on the surface of a lagoon and these chelators form on it. Phosphate in fairly high concentrations might be released. With this scheme, it is possible to get an excess of phosphates over calcium.

SIEVER: Only if there is a high bicarbonate situation. If the pH moves up around 9 the calcium will start depleting. The only mechanism I can think of to deplete the calcium is to get an excess of phosphate. Other than that the answer is really no.

ABELSON: This afternoon a certain amount of floundering occurred in part because, perhaps, we didn't have all the knowledge that is available, but also, perhaps, because it simply hasn't been accumulated.

In the work on the origin of life we should try, one way or another, to get as many solid facts as we can. For example I know that Dr. Barghoorn has been interested in the inventory of carbonate rock. It might be desirable to encourage some geologists to go out and make a survey of what the geological column is like in the earliest rocks, so that in a subsequent discussion we would have more solid information on this matter. We also need an inventory of reduced carbon.

Another item mentioned earlier in the day was the matter of adsorption of organic matter on clays. Everybody skirted around it. Some work has been done on it. Incidentally, I have observed that arginine is quite highly adsorbed on clay. Some people should really do something solid on problems of adsorption on clay-type materials.

Later this afternoon we discussed the temperature in the earliest Precambrian. It would be good to know the full status of evidence for low temperatures, glaciation, and so forth.

You might recruit some geologist who wanted to go rambling around the world anyway by telling him how important it would be to have definitive statements on evidence for early glaciation. Perhaps you could even talk the NSF into supporting this man.

One final comment relates to this matter of ammonium. There ought to be evidence for it in the very old granites. Ammonium which is a pretty tough, stable ion, can proxy for potassium. It ought to be possible to find the ammonium in ancient rocks.

Although I don't think he studied the oldest, I believe Rayleigh (1939) did find ammonium in some pretty old rocks. You would have to be careful to see that you weren't collecting a specimen next to a graphite schist, where obviously the ammonia had been cooked out of organic matter. The occurrence of this might be such that it was evidently deposited as a result of weathered clay having been in contact with the ammonium in sea water.

I may be beyond my depth in this suggestion, but certainly it would be desirable to seek whatever evidence one could from the earth for this kind of hypothesis.

Thank you.

ORGANIC MOLECULES IN INTERSTELLAR SPACE
Friday Morning Session

YOUNG: Today's topic, "Organic Molecules in Interstellar Space" has recently emerged as something of interest as well as potential relevance to the question of the origin of life. This morning we will talk about how potential or direct the relevance is. Our discussion leader, Toby Owen, will introduce Dave Buhl from Greenbank and other conferees knowledgeable in this area. So I suggest we just get going.

OWEN: I will just bridge the gap between Dick Young and the people really knowledgeable about organic molecules in space.

The problem is how this complicated zoo of molecules turning up in interstellar space gave us the earth and its early environment. A question we have to approach, for example, is whether the molecules seen in interstellar space would be available on the solid earth.

This obviously relates to what occurred during the early accretion phase of the earth. Put another way: at what stage in the earth's formation did the material forming the earth lose the volatiles that we know have been lost from studies of the noble gases? What was the near earth environment? What effect did the sun have during that time?

Unless we attack those very difficult questions, we really won't be able to evaluate the significance of the interstellar molecules.

Planetary astronomers have approached some questions about conditions on the earth at the very early stages by comparison with other planets. Concerning volatiles, the outer solar system presents us with some of the best evidence we have. In particular, Jupiter and Saturn seem to have these elemental abundances in their atmospheres very similar to those existing in the sun. This suggests these giant planets have retained material from the original cloud of gas and dust − the "primordial solar nebula" from the time that these planets consensed out of the nebula. Therefore, we might seek evidence of complex organic molecules of the kind being found in interstellar space on Jupiter and Saturn.

On the other hand, because Jupiter and Saturn are very rich in molecular hydrogen and are fairly high density environments they are probably not good places to look. One wouldn't expect necessarily that these molecules would survive.

A much better place to look for retention of complex organic molecules might be on the satellites of the outer planets, where conditions probably have remained at low temperatures throughout the history of the solar system, yet the gravitation is sufficiently low so that the excess hydrogen would have escaped. In fact, a Saturn satellite, Titan, now has an atmosphere of methane, and a reddish color. We certainly should know if there are complicated organic molecules associated with the surface and possibly even in the atmosphere. At the present time in the solar system the best place of all to look for these molecules is probably the comets.

SAGAN: Toby, is your feeling that Jupiter is a less favorable place to look for organics because of the excess of hydrogen? Everything will react back to methane, ammonia, and so on, so there might be some steady state concentration of organics, but no survival of primordial organics? And since on the satellites of the Jovian planets there is not an excess of hydrogen, so you might there expect preferential preservation.

OWEN: Yes.

SAGAN: But there is a question of generation. If the present production rate is very large, then the primordial stuff will be swamped out.

All four Galilean satellites of Jupiter, plus Jupiter 5, live in the radiation belt, the massive Van Allen belt of Jupiter, and so high energy charged particles are constantly entering the atmosphere of these satellites if they have atmosphere. This is very much like an electrical discharge in such an atmosphere, and I would suspect that organic molecules are being produced all the time there. Thus, any primordial residuum may be swamped by contemporary production.

OWEN: Yes. I should have been more specific when I said satellites of the outer planets. I would exclude the Jupiter satellites which show no evidence of atmospheres, of organic material, and are relatively close to the sun (Jupiter being the closest of the outer planets). A possible exception is the reddish coloration of Io, which might be explained that way.

SAGAN: They are all reddish.

OWEN: Io exceptionally so. But when I say satellites of the outer planets, I mean Saturn and farther out, and the farther out the better.

But comets I think are really even more interesting. Long before water or ammonia was detected in interstellar space, the molecules were expected from fragments that showed up in the spectra of comets (Table 9). The first four items, CN, CH, OH, and NH, showed up in interstellar space as well. It was speculated that these radicals which appeared in the comets and their tails (the last five ions are observed in the tails of comets) were coming from parent molecules in ices which really formed the body of the comet itself.

The ices would be solid ammonia, water ice, and perhaps a methane hydrate. Some adventuresome people suggested that maybe acetylene or HCN was present, because obviously there is some difficulty getting C_2, C_3, and CN from methane.

TABLE 9. MOLECULES IDENTIFIED IN COMETARY SPECTRA

Coma:	CN	C_3
	CH	C_2
	OH	NH_2
	NH	

Tail:	CO+
	N_2^+
	OH+
	CO_2^+
	CH+

Now, with all the new features discovered in the interstellar medium, apparently molecules are available, or should be available in the cloud which condensed to form the solar nebulae which, in fact, have C_3 or C_2 in them, e.g., cyanoacetylene. The comets may have been the first objects that condensed out of this collapsing cloud, and may thus represent a sort of deep freeze sample of the material that was available for planet formation from the very beginning, preserved for us because of their great mean distance, thousands of astronomical units, from the sun.

SAGAN: The condensation of globular clusters as the galaxy forms provide a very nice stellar analogy to this picture.

OWEN: Yes. The galaxy, like the solar system, has a very flat distribution, but the globular clusters exhibit a spherical distribution around it. This is true of comets. All the planets are more or less in the same plane, but the long period comets, the ones referred to here, show a spherical distribution also.

SAGAN: We also know the globular clusters are old from their turnoff from the Hertzsprung-Russell diagram.

OWEN: Yes. The idea is that the globular clusters would have condensed out of the total pregalactic cloud — earlier than the stuff that went in to form the galactic plane, analogous to the suggestion of what happened in the case of comets. This also avoids the problem of how comets were made in the plane of the galaxy, which has been very bothersome. There is not enough density at distances from the sun far enough to preserve these kinds of ices in these small bodies. But if they are made by condensation of stuff that is moving through the cloud in the very early stages, they might be able to grow in that way, presumably in the way that the globular clusters formed.

Enough general background. Today first we should ask what molecules have been observed in the interstellar medium. How are they distributed around the galaxy? Is there a tendency for them to be concentrated in clouds, or are they uniformly distributed? What molecules have been looked for and not found? What other molecules do you suggest should be looked for? Phosphorus and sulfur compounds might be high on that list. Finally, what would we expect the history of these molecules to be during the formation phase of the solar system – specifically, in the neighborhood of the earth? What happens to the molecules as the planets begin to form?

With that background, I turn the discussion over to Dave Buhl, who can talk about what is really happening out there.

Radiotelescopes and Molecular Identification in the Interstellar Medium

BUHL: This is a message brought to you from one hundred to one hundred thousand light years away. I will start by presenting our equipment, types of telescopes, and spectral line data – essentially the methods we use to detect these molecules, and the sort of reliability we place upon the identity of a spectral line.

Figure 19 shows the 140 ft. telescope in Greenbank. It has been used for about six of the molecule detections which I'll discuss. It is good at wavelengths between about 2 cm and 36 cm, which is the longest wavelength that has been detected so far.

Figure 20 shows the 36 ft. telescope, smaller by a factor of 4, but with a surface such that we can use it down to about 1 mm in wavelength. In radioastronomy we are limited by diffraction, and the resolution of these telescopes is given by the wavelength divided by the diameter. Because this telescope is used at much shorter wavelengths our resolution is much better. The 36 ft. telescope was used for about a dozen of the molecular lines, such as hydrogen cyanide and carbon monoxide, which were discovered in the millimeter wave range. CO is a particularly interesting molecule, because it seems to be very stable and very widely spread. Its densities are much more abundant than any of the other molecules which I will be discussing.

Figure 21 gives you an idea of our resolution relative to an optical photograph of the Orion nebulae. This is a relatively nearby object, as the clouds go, approximately 1500 light years away. The distance between the sun, which is about two-thirds of the way out in the galactic disk, and the galactic center is about 30,000 light years. So this cloud is less than a tenth of the distance away. The Orion nebula happens to be located in the anticenter direction, another direction, another 1500 light years further out than we are. It is a very nearby object in which we believe star condensation is occurring. It is a very rich molecular source, probably one of the most important sources of molecules in the sky. The excitation conditions are such that we see a great variety of molecules in this particular nebula.

Figure 19. Radiotelescope, 140 ft. dish, of the National Radio Astronomy Observatory, Greenbank, West Virginia.

Behind the bright continuum is a very large region of dust and obscuration. We will discuss the question of dust, because it is closely interrelated with the gas in these clouds. Apparently dust is very important for the generation and protection of these molecules. The details of this process are not well understood. This very difficult problem is probably as closely associated with the prebiotic problem as any problem in astronomy. We are trying to understand the generation of similar molecules, in somewhat different environments, and are just beginning to study how these molecules are formed.

MILLER: What is the diameter of the nebula?

BUHL: The diameter of this region (OH resolution bar, Fig. 21) is about ten light years across, or 10^{19} cm. The astronomical scales are so large that we usually work in powers of ten, with units such as light years or astronomical units (AU). One AU is the distance between the earth and the sun, approximately 8 light minutes.

Figure 20. Radiotelescope, 36 ft. dish of the National Radio Astronomy Observatory at Kitt Peak.

We have a number of quite different sized clouds here. Looking at the carbon monoxide line, the cloud in this nebula is larger than the entire photograph; so the carbon monoxide extends at least 15 light years, much further than the illuminated, bright regions in this nebulae. The Orion nebulae is a very complex object. Apparently star generation is occurring in the central part. At least one infrared object is known. An infrared star is assumed to be a star just beginning to turn on, essentially a protostar or very young sun.

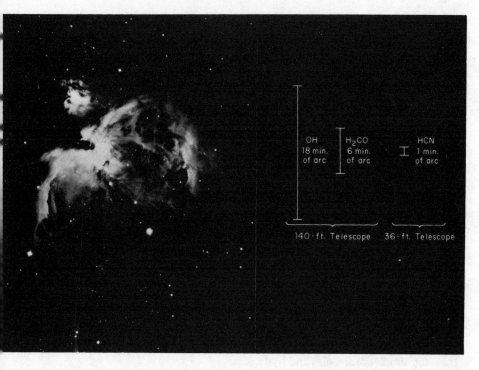

Figure 21. Orion nebula: Comparison of optical resolution (left) with resolution obtained with the 140 ft. radiotelescope (Fig. 19) and the 36 ft. radiotelescope (Fig. 20).

Roughly coincident with the infrared star and infrared nebula are water and OH masering sources. I won't discuss this in detail, because it is not very relevant to generation of prebiological molecules in space, but several molecules in the interstellar medium (e.g. OH and water) exhibit a peculiar masering phenomenon. It is essentially equivalent to a laboratory laser, except that there is no cavity or mirrors at the end to make it into an oscillator. It is a very high gain amplifier on a particular molecular line. The details are not understood even by astronomers who study this, but it is established that there are enormous quantities of energy involved in these sources. For example, the amount of energy coming out on one of the water lines in the microwave region is nearly equal to the total energy radiated by the sun at all wavelengths. Yet some process funnels this energy down into one single microwave line. Tremendous amounts of energy are involved in this process; we think that this may be a mechanism by which a cloud cools when it is contracting. It starts out at several light years in diameter and slowly contracts until it is roughly solar system size. During this period some cooling of the cloud is required in order to make it contract. This masering line may be an example of this process but the details are not well known.

Now with respect to sizes of the molecular clouds in Orion: the CO cloud is about 15 light years in diameter. The hydrogen cyanide in this source is roughly eight minutes of arc or about 4 light years in diameter. The masering sources, such as the water or OH lines, are more like an astronomical unit in size (light minutes rather than light years). They are much smaller objects with solar system type dimensions implying that the maser, as well as being very intense, takes place in a very small region.

OWEN: What is the density of the stuff you are looking at?

BUHL: Marvin Litvak says it is up to 10^8 per cc.

SAGAN: For which molecule?

BUHL: For the H_2O molecule.

OWEN: Is this based on one astronomical unit?

BUHL: Yes, that value is strictly by implication; the density required for pumping the maser.

OWEN: Might that not be a model-dependent kind of number?

BUHL: Yes. The numbers which I'll put down a little later are more like 10^{-4} per cc for some of the most complex molecules, and 10^4 per cc for the simpler molecules and hydrogen.

SAGAN: There is a progression of size from OH and H_2O. OH and H_2O are the smallest angular diameters, HCN and then CO are larger. I don't know how HCN and CO compare, but does this correspond to radiation stability?

BUHL: It is due to two things: either the stability of the molecule to UV radiation or the density required to excite the molecule. The fact that CO is an extremely stable molecule fits in with the larger size of the cloud. The size may also be related to the excitation conditions. Certain molecular lines are excited into emission by some radiation or collision process above the background continuum. This is a density dependent effect. Since there is probably a gradient in density as one moves out away from the central region of the nebulae, molecules which have a very short lifetime, like hydrogen cyanide, require high densities and will only be excited over a small region. Molecules like CO have very long lifetimes and will be excited over a much larger region. It is a little difficult to unscramble but it is probably some combination of the two effects.

It is possible that we are being fooled by a projection effect in Orion. The bright nebula, a very hot region, may be in the foreground and probably does not contain molecules. These are called H II regions, in astronomy we use the Roman numberal II to identify an ionized region (HII is equivalent to H^+). They are very hot (about 10,000°K) so these are regions where molecules would not be expected to stay around very long. There are generally excited by some central star, the UV light from the star doing the ionizing and some involve very high energy particles, which again would be very inhospitable for molecules.

The molecule cloud which we are seeing in this particular source may be in the large dust cloud behind the emission nebulae. The dust in question is a very elusive ingredient. We can't do direct spectroscopy on the dust, nevertheless we have a good idea of what the composition of the dust is. The models suggest it is probably a core of silicates, graphite or carbon, surrounded by some mantle which may be an icy mixture of various molecules. Essentially, it is a scaled-down version of a comet, with dimensions of microns, rather than kilometers. These are very small particles. We know the particles are there because they obscure stars and redden the star light. When one plots the intensity of some stars versus wavelength, one can see the reddening due to the particles. Hence the evidence for the dust is more by implication than by direct spectroscopic measurements. The characterization of the dust at present is not as good as our knowledge of the gas. The nature of the dust may very well be revealed by the kind of molecules that are produced in the dust clouds.

Table 10 shows the molecules presently identified in the interstellar medium. We have a list now of 21 molecules, which have been detected by radio telescopes and 3 by optical telescopes including a recent detection of molecular hydrogen. I'll put these down including the optical molecule even though it is difficult to make comparisons between optical and microwave results. It turns out in the optical results that if one has too much dust one cannot see the star at all so one must use certain stars which are generally not located in the densest part of the nebulae. Hence it is difficult to compare densities and properties of optical and radio data.

In table 11 we have a column indicating the number of lines, to give some idea of reliability of a particular identification and another factor, the peak intensity, which gives an indication of the certainty of detection. With the present number of molecules and number of lines we have something like 50 transitions which have presently been found in about two decades of frequency range, between about 1.5 and 150 GHz. The chances of two of these lines overlapping within roughly 300 km per second is about one per cent. This gives a number like 99 per cent reliability for these identifications, given the present known constituents. Depending on the complexity of the interstellar medium, this number will change somewhat. However, we are approaching the limit of sensitivity for the larger molecules with the telescopes and the receivers we have at present.

The line called X-ogen appeared in the spectra during a search for the carbon[13] isotope of HCN. It has not been identified to any satisfactory degree, although there are a number of suggestions. HCO^+ and CCH are two possibilities.

TABLE 10

MOLECULES FOUND IN THE INTERSTELLAR MEDIUM

Year	Molecule	Symbol	Wavelength	Telescope	Initial Discovery
1937	–	CH	4300 Å	Mt. Wilson 100 in	Dunham
1940	Cyanogen	CN	3875 Å	Mt. Wilson 100 in	Adams Mt. Wilson
1941	–	CH$^+$	3745-4233 Å	Mt. Wilson 100 in	Adams
1963	Hydroxyl	OH	18, 6.3, 5.0, and 2.2 cm	Lincoln Lab 84 ft	MIT/Lincoln Lab
1968	Ammonia	NH$_3$	1.3 cm	Hat Creek 20 ft	Berkeley
1968	Water	H$_2$O	1.4 cm	Hat Creek 20 ft	Berkeley
1969	Formaldehyde	H$_2$CO	6.2, 2.1, 1 cm 2.1, 2.0 mm	NRAO 140 ft NRAO 36 ft	U Va/NRAO/U Md/U Ch
1970	Carbon monoxide	CO	2.6 mm	NRAO 36 ft	Bell Labs
1970	Cyanogen	CN	2.6 mm	NRAO 36 ft	Bell Labs
1970	Hydrogen	H$_2$	1100 Å	UV Rocket Camera	NRL
1970	Hydrogen cyanide	HCN	3.4 mm	NRAO 36 ft	NRAO/U Va
1970	X-ogen	?	3.4 mm	NRAO 36 ft	NRAO/U Va
1970	Cyanoacetylene	HC$_3$N	3.3 cm	NRAO 140 ft	NRAO
1970	Methyl alcohol	CH$_3$OH	36, 1 cm, 3 mm	NRAO 140 ft	HARVARD
1970	Formic acid	CHOOH	18 cm	NRAO 140 ft	U Md/Harvard
1971	Carbon Monosulphide	CS	2.0 mm	NRAO 36 ft	Bell Labs/Columbia
1971	Formamide	NH$_2$CHO	6.5 cm	NRAO 140 ft	U Illinois
1971	Silicon oxide	SiO	2.3 mm	NRAO 36 ft	Bell Labs/Columbia
1971	Carbonylsulfide	OCS	2.7 mm	NRAO 36 ft	Bell Labs/Columbia
1971	Acetonitrile	CH$_3$CN	2.7 mm	NRAO 36 ft	Bell Labs/Columbia
1971	Isocyanic acid	HNCO	3.4 mm, 1.4 cm	NRAO 36 ft	NRAO/U Va
1971	Hydrogen Iso-cyanide	HNC	3.3 mm	NRAO 36 ft	NRAO/U Va
1971	Methylacetylene	CH$_3$C$_2$H	3.5 mm	NRAO 36 ft	NRAO/U Va
1971	Acetaldehyde	CH$_3$CHO	28 cm	NRAO 140 ft	Harvard
1971	Thioformaldehyde	H$_2$CS	9.5 cm	Parkes 210 ft	CSIRO Australia

TABLE 11. DETECTION AND IDENTIFICATION
RELIABILITY FOR INTERSTELLAR MOLECULES

Molecule	Line Intensity	Number of Lines
H_2, CH, CN	Optical	Several
OH	$200°$K	8
CO	50	3
CN	1	2
CS	5	1
SiO	0.5	1
H_2O	2000	1
HCN	10	2
X-ogen	5	1
HNC	1	1*
OCS	1	1
NH_3	5	4
H_2CO	5	8
HNCO	2	2
HC_3N	1	2
HCOOH	0.05	1
CH_3OH	2	6
CH_3CN	1	4
NH_2CHO	0.2	3
CH_3C_2H	0.2	1

*Calculated line frequency

The number of transitions of OH includes the number of excited states and the O^{18} detection. Many of these lines are split exhibiting more than just a single line. I have tried to account for that too in these numbers.

BUHL: The identification of formic acid is rather tentative. It is a very low signal level, and lies in the middle of some O^{18}H lines. The line is in an extremely crowded region of the spectrum and the identification is up in the air at the moment.

The large intensities, OH and water, are due to the masering lines, these are extremely excited molecules.

ORGEL: For those of us not in the field, could you explain what your temperatures mean?

BUHL: A line temperature of $200°$K is related to the equivalent temperature which would be measured if a black body were placed in front of the telescope and heated up to $200°$K. We use temperatures in radioastronomy because the Planck curve is linear in temperature for almost the entire spectral region in which we work. Rather than using microwatts per square centimeter per Hertz, it's much more convenient to think in terms of temperature.

Our noise levels are from one degree Kelvin for the millimeter wave lines to as low as a hundredth of a degree for the centimeter wave lines. If you think of this as the noise, you will get some idea of the reliability of a detection.

ORGEL: This is a measure of reliability, but only indirectly of the actual identity.

BUHL: Yes. Think of the intensity as a measure of reliability in terms of whether or not the line is there. The number of lines is more a measure of identity — knowing that the line is there, how certain you are that it is that species and not some other molecule.

OWEN: What would be the reliability of those temperatures?

BUHL: We are down to 0.04° in formic acid which is questionable. The others are quite certain. We have a number of different sources, and the velocities match up, giving us other ways of ascertaining what we are looking at.

But this is not the only way of going from the energy level diagram (the known microwave transitions) directly to the molecule. There are splittings and isotope lines to aid the identification. I feel for a single line that the reliability is about 90 to 99 per cent, depending on how crowded the spectrum is. With two or three lines the certainty is much higher.

ORGEL: Is it generally true that once you know the molecule, you know what line to expect, if they are all there?

BUHL: No. This is a problem, particularly on more complex molecules. We know only that the lower levels are populated, because the temperatures are very low. So only the lines which are close to the ground state of the molecule are normally seen. If we do not have some indication of what energy level the line originated, the laboratory spectra taken at room temperature are not always very useful. With all of the molecules that have been found up to now we have a general idea of what the energy level structure looks like.

The three diagrams for ammonia. OH, and formaldehyde are shown in Fig. 22. These show only the very lower states. For those who don't like wave numbers, which may be a small minority, this can be converted roughly to temperature. It requires ten degrees of excitation to populate the lowest transition of formaldehyde that has been found. In the interstellar medium temperatures are from 10 to 100° K, so we're talking about excitations of these kinds of levels. Occasionally something odd like the water transition shows up at a level of 446 cm^{-1}. That is a very peculiar anomaly, and requires a very dense, small region in order to get the excitation of that kind of a transition.

The OH levels are split into four lines, which are quite distinct. The first radio detection of a molecule was OH which was first discovered in absorption by Weinreb, Barrett, Meeks, and Henry (1963). About two years later a number of very intense emission sources were found by Harold Weaver's group at Berkeley. The line ratios were so enormously peculiar that they originally called it mysterium. This confusion came about because the spectroscopic

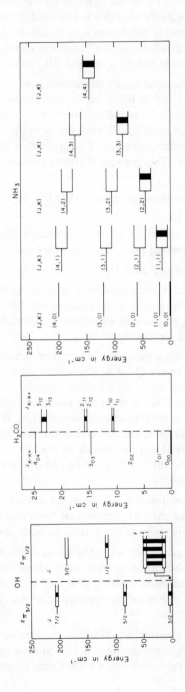

Figure 22. Energy level diagram of hydroxyl, formaldehyde, and ammonia (only the very lower states).

identifications were thought to hinge as much on the intensities as on the line frequencies. We are now fairly sure this is an effect of masering, which unbalances the levels producing conditions very much out of thermodynamic equilibrium. In the interstellar medium things are very seldom in thermodynamic equilibrium. They are normally out and it is just a question of the degree.

In the case of formaldehyde, six transitions have been found — the three indicated in Fig. 22 and another three due to transitions between the $J = 1$ and $J = 2$ levels of the molecule. I have added another two for the carbon[13] and oxygen[18] isotopes which have been identified. Another way of identifying a molecule is to find an isotopic species. The problem is that the isotope ratios are off by factors of ten. This is a very difficult problem, which has yet to be unscrambled.

OWEN: Could you please explain: do the C^{12} to C^{13} ratios vary at different places in the sky?

BUHL: Yes. In some places there is apparently 90 to 1, C^{12} to C^{13}, roughly the normal terrestrial ratio, while in other places it is 5 or 10 to 1.

OWEN: It tends to approach the ratios found in carbon stars?

BUHL: Right. That is one way of looking at it. Another is that the carbon[12] line of the more abundant species is very saturated. By saturating a cloud enough these ratios can go from 90 to 1 down to 10 or 5 to 1, but large optical depths or E-folding factors are needed.

In ammonia there are a number of transitions which are found in emission. The molecule is heated to about $50°K$ above the background. In the case of formaldehyde the molecule is in equilibrium with the $3°K$ background and is sometimes cooled below $3°K$. When a radio source is observed, the telescope receives radiation, part of which has been absorbed by the molecules in the clouds at the formaldehyde frequence so that the spectrum seen is an absorption line. The frequency is shifted because of the velocity of the cloud with respect to the earth. The shift in this line can be as much as 0.1% of the rest frequency of the molecule.

After studying several molecules, we begin to know the characteristic velocity of the cloud and then we can convert from one molecule to another to determine how closely they match in their velocity spectra.

In the case of formaldehyde when the telescope is pointed off the radio source no emission or absorption spectra can be detected implying that the cloud is roughly in equilibrium with the $3°K$ background temperature. The excitation temperature describes the population of the two levels producing the line.

Because of the nonthermal equilibrium, the kinetic temperatures in these regions may be much higher than the $3°K$ background temperature, and the associated velocities in turn even much higher than suggested by the kinetic temperatures. There are several processes here to which different temperatures could be assigned, so that this is a very nonequilibrium type of arrangement.

OWEN: Is your resolution good enough to give you any feeling for differential motions within the cloud?

BUHL: I'll indicate where there does seem to be some structure. Just the width of the line is an indication of the motions within the cloud. Figure 23 shows a particularly complex spectrum; the absorption due to formaldehyde and hydrogen cyanide emission in the path between the radio-telescope and the galactic center, about 30,000 light years away. These various absorption features are not structure due to the molecule, but velocity structure. So instead of plotting this as a function of frequency, wave number or wavelength, we plot it as a function of velocity.

The large feature at +40 km/s is thought to be directly associated with the galactic center. The zero velocity feature is relatively local formaldehyde. The others are thought to be other spiral arms which happen to be in the path of the radiation. We think formaldehyde (and also CO and OH, which have similar spectra) is a widely distributed molecule permeating most of the galactic spiral arms.

ORGEL: With H and CO do each of the lines split into a velocity pattern equivalent to this?

BUHL: Yes, but the lines are weaker, and the relative intensities are different. Most other molecules just seem to pick up the +40 km/sec feature.

The formaldehyde lines are widely separated, one at 6 cm and the next line up at 2 cm, a factor of 3 in wavelength. Detecting these involves mounting a different receiver on the telescope, and doing an entirely different experiment, not just tuning a spectrometer. The noise for the formaldehyde spectrum is down around the tenth of a degree level for roughly one hour of integration time. The hydrogen cyanide line has about one degree of noise at a wavelength of 3 mm. For such a short wavelength receivers are much more noisy and temperamental. We do not have the signal-noise necessary to pick up some of the other velocity features in the hydrogen cyanide line.

SAGAN: Isn't there also a local HCN feature, or might that be noise?

BUHL: The signal is not big enough to say whether they are real or not. We have been unable to average for more than 15 minutes, and the receiver is too unstable.

ORGEL: Does "local feature" mean somewhere near here, a feature without a Doppler effect?

BUHL: Yes, the local spiral arm in which the sun is sitting. "Spiral arm" is a vague term because only some galaxies have very distinct arms, in others

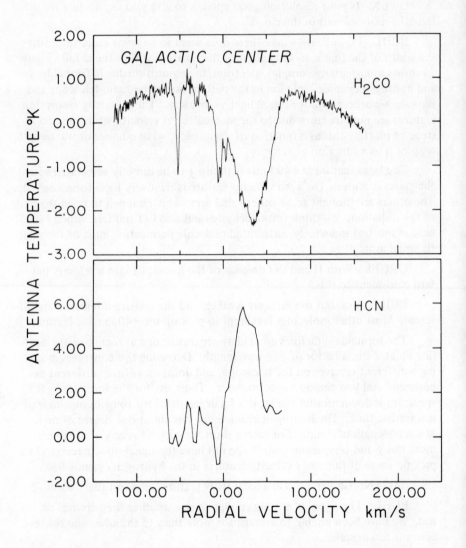

Figure 23. Radiospectrum of the absorption due to the emission of formaldehyde and hydrogen cyanide.

they are very diffuse. A problem with which you probably are not familiar, is that in optical astronomy you can see only a very short way into our galaxy, not even as far as this particular source near the galactic center because of the obscuring dust. We have no optical photographs of some of these features here. Only in nearby clouds, like the Orion nebula, have we an idea of what the optical structure looks like. Such nearby nebulae are interesting because we are convinced that the dust is related to the molecules.

A brief table for the interstellar environment contains numbers very different from what you are used to in the laboratory (Table 12). A radiation temperature of roughly 3°K is presumably the remnant of the primordial fireball. The peak of 3°K radiation is about 1 mm in wavelength. Since it is a thermal distribution, the radiation is not effective in the infrared, but in the microwave range this is an important contribution to the excitation that these molecules are bathed in. Part of the environment involves a collision temper-

TABLE 12. INTERSTELLAR ENVIRONMENT

Radiation Temperature	$3°K$
Grain Temperature	$20°K$
Dust Cloud Kinetic Temperature	$10°K$
Normal Cloud Kinetic Temperature	$100°K$
Thermal Velocities	0.5 km/s
Turbulent Velocities	10 km/s
Cloud Lifetime	10^7 years
Cloud Size	10 light years (10^{19} cm)
Cloud Mass	100 solar masses
Cloud Distance	100-100,000 light years
Density of H_2	10^2 cm^{-3}
Density of all Molecules	10^{-1} cm^{-3}
Density of Dust/Hydrogen	10^{-12}
Grain Size	10^{-5} cm
Grain Mass	10^{-14} grams
Mass of Dust/Hydrogen	10^{-3}
Molecule Collision Time	1 year
Density of H_2 Required to Excite Molecules	10^2-10^8 cm^{-3}

ature of about 100°K which represents the thermal velocity at which these molecules are moving. Because interactions are so seldom (usually years) there is no thermal equilibrium between these two. So you can come up with a number of different temperatures.

The velocity widths of these features indicates that the kind of turbulent motion involved is about 10 km per second. The temperatures of 100°K gives you motions within the cloud of about a tenth of a kilometer per second just due to the thermal motion of individual molecules. So we know there are over-all large scale motions within these clouds. Any reasonable interpretation of most of these molecules leads to densities of 10^{-2} to 10^{-4}/cc, so except for hydrogen and CO the molecule densities are much less than 1 per cubic centimeter.

SAGAN: Except in clumps.

BUHL: Yes. We may discuss the thickness of clouds later. I think these clouds probably are not as uniform and homogeneous as they seem. The numbers may be higher in very small sources, which I would refer to less as a cloud than as a protostar condensation formation. The numbers Marvin Litvak needs are more like 10^8/cc, certainly not the density of an average cloud.

The diameters are about ten light years. These objects are from 100 to 100,000 light years away from us, so essentially they are distributed pretty well across the galaxy. In most cases they are local condensations, and the question of how uniformly distributed these molecules are is still open. They are apparently associated with dust, which itself seems associated with places where star formation takes place. The cycle of the birth of stars is somehow interrelated both with these clouds and with the process of molecule formation.

The time scale of the collision time between clouds is about ten million years. This may be the time scale of the lifetime of the molecules but it is also possible that they survive the collapse and accretion phase too. The question of the survival time of these molecules seems to be somewhat open.

OWEN: Your list of molecules is presumably derived from observations made in many different directions, at different frequencies, and so on. Which molecules tend to be associated in a given cloud? Is there any pattern emerging?

BUHL: It is not that. They all seem to be in the same clouds, but the trouble with very weak lines is that they are only picked up in the galactic center, where the densities are believed to be the highest. If the density falls off much below that of present molecules they are not going to be detected because the signal is below the noise threshold of the receiver. For example, HCN is roughly the same strength as formaldehyde, but the noise is about ten times worse, because the receiver for HCN is not as good as for formaldehyde. There are roughly ten times as many sources for formaldehyde as for HCN, hence if we had comparable sensitivity we probably would pick up an equal number of sources for HCN. There is a very high coincidence between all of the molecules for which we have reasonable statistics.

OWEN: Do you mean that if you find a cloud that has a lot of one of these molecules, you tend to find all of them?

BUHL: Yes, but they may appear at slightly different velocities, which may mean some region of the cloud is more favorable for one than another. But we don't believe that there are clouds exclusively OH in one place and formaldehyde in another. We think there is some mixing. Grains may be accumulating this goo, picking up these various ices and then being blasted off by shock waves or by high energy particles, or somehow producing a variety of molecules, just depending on which chunk of this grain gets knocked off.

There seems to be no selection effect favoring one molecule over another except on the basis of differential survival in the UV environment. Apparently, a fair amount of dust is located in these clouds, and the only stars are very young infrared objects. Outside this cloud may be a number of stars which are blasting UV photons at the exterior of the cloud, which only get part way in before they are absorbed or scattered by the dust particles. The more hardy molecules, presumably survive to greater distances away from this cloud. This may be the implication of Carl's remarks about the Orion nebula (CO has a much greater UV lifetime), where CO is able to push further out into the unprotected region than hydrogen cyanide or formaldehyde might be. We think the densities we measure tell us something about the formation and destruction of these molecules, something about the chemistry going on.

The cyanoacetylene line in the galactic center is on fairly firm grounds since it has a second split component of the line and the noise level is down around a tenth of a degree.

ORGEL: How can you tell whether you are dealing with cyanoacetylene or with another line in the same general region which is velocity shifting?

BUHL: This is what I was trying to get to but I didn't write it down. If you know the rest frequency of the line, and if it can be anywhere within 30 km of the center frequency you have an error box of 1 part in 10^3. We have about 50 transitions spread out over 1.5 to 150 GHz and it turns out 5,000 of these boxes can be fit along this line. Hence there is one known line for every 100 boxes along the spectrum. Thus there is a one per cent chance that these will overlap. Another point is that if a line appears at an unusual velocity while all the molecules ever seen in this cloud were at another velocity 100 km away you would be very suspicious. Of course, I have only counted those transitions that we know now, and there must be others we'll find when our sensitivity is increased. As the spectrum becomes more crowded there's a problem of misidentifying a line.

ORGEL: I now understand how you distinguish your transitions from those of other molecules you have already found. But as a chemist, I wonder: How can you be sure that it is a certain new molecule. How can it be distinguished from a much larger family of molecules that haven't been found?

BUHL: As I mentioned to Carl yesterday about formaldehyde, most people would suggest many possible organic molecules — but in the beginning astronomers were reluctant to accept formaldehyde because it was too big. At the level we are at now, we are running into sensitivity problems with molecules of more than two non-hydrogen atoms.

We ask: How many molecules with three non-hydrogen atoms are abundant enough to be considered likely? Possibly S, C, N and O, and maybe a few others are abundant enough to produce molecules. The class is certainly not bigger by a factor of ten than what I have demonstrated. Carl, do you object?

SAGAN: For the sake of argument I do. One appealing model of the origin of these molecules particularly far from dense sources is that the interstellar grains are much more complex organic molecules than those radiation degradation fragments you have drawn.

If so, there may be much more complex molecules than yours. Some might have large numbers of lines. The real list may be ten or 100 times larger, just in terms of the number of molecules. If so, Dave, then your 90 per cent probability goes way down.

BUHL: I agree, Carl, but at the moment we are sensitivity limited. I think the whole pyramid is filled up. There may be amino acids, and even amoebas out there! But presumably the density is way down.

SAGAN: I'm saying maybe not. Maybe there are a larger number of complex molecules. The small ones here may be fragmentation products constantly regenerated by larger molecules which live for a brief time until destroyed by the radiation field.

BUHL: I hope you're right, because that will keep us in business longer!

SAGAN: In the cases where there are several individual lines, you are okay. But in other cases, where there is only one line, I think Leslie's uneasiness is valid.

BUHL: I agree we might be suspicious about the 1's, especially since we don't know what X-ogen is anyway!

BADA: What would be the source of more complex organic material?

BUHL: The problem is that obviously a grain is a molecule. So Carl is right. Big molecules are out there, but their density isn't sufficient for us to pick them up.

Another problem creeps in. The bigger the molecule, the more these transitions are compressed which pushes the frequency way down and spreads the energy out over a whole lot of lines. I suspect a factor of ten increase in our sensitivity might produce a factor of ten or a hundred more molecules.

If there are a lot of lines, if the spectrum were crowded we'd be picking them up, because we do searches over frequency ranges, looking at more of a region than just one individual line might indicate. In terms of confidence level, I would agree with 90 per cent on a single line, but I would not agree with a drop to 50 per cent.

ORGEL: From the amount of the spectrum already carefully scanned, can you estimate how many new lines of comparable intensity might be there, but are as yet undiscovered?

BUHL: We haven't scanned really large ranges. Certainly, no more than ten per cent of the radio spectrum has been gone over even crudely, because it is very difficult to scan two decades of frequency. Essentially there are 5,000 of these boxes, and if you must spend a day on each of them, that's a lot of boxes, and eventually these boxes become literal receiver boxes that have to be dismounted and mounted on the telescope.

It is the same problem as with GC-MS. Not everybody has a telescope in his back yard, and the demand for these instruments is very high. We might get two or three weeks a year of telescope time.

We have to figure what the point of diminishing returns is, and the approach that Lew Snyder and I have used is to take known transitions as a first order attempt to search for . . .

OWEN: Are you talking about molecules you have sought and not found?

ORGEL: And also ones found by accident?

BUHL: X-ogen was an accident, and formic acid is one we looked for and didn't find.

USHER: What was that?

LEMMON: Formic acid, which has your lowest reliability index?

ORGEL: Are there many lines you have found without especially setting up to look for them?

BUHL: No, only X-ogen. In the optical range there are something like 25 diffuse bands which have been known for about thirty years, some more recently than others. Nobody has yet identified a molecule except Fred Johnson, whose identification of prophyrin (Johnson, 1970) is somewhat dubious. If enough atoms are put together in a certain order, you are bound to hit a certain number of these.

USHER: It is interesting that molecules that are most polar and more likely to self-replicate are the ones —

BUHL: What do you mean by polar?

USHER: Formic acid, for instance, could easily aggregate into something that makes it less likely to be seen toward the outer edge of the cloud.

MILLER: Dimer can not be formed at these pressures.

USHER: I don't know. If there are particles —

NAGYVARY: If a molecule has limited rotational freedom and fewer lines, does it help the identification? Is the linear spatial arrangement of cyanoacetylene related to the accuracy of the assignment?

BUHL: Yes, because the energy levels of some of these simpler linear molecules, like HCN, are simple linearly increasing energy differences. For instance, in the case of CS, because they happened to have a receiver up for the 150 GHz region, they looked for the J = 3 to 2 transition, whereas there is also a 2 to 1 at 100 GHz and a 1 to 0 at 50 GHz. For a nice molecule like CS or HCN there is a nice simple arrangement. This gets more complicated in a case like formaldehyde, which is still much simpler than amino acids where the energy levels are very difficult to calculate. There has been some attempt to do laboratory measurements, but the problem is that the energy structure of some of these molecules is not really known. For OH there was a calculation of the frequency, and then a laboratory measurement searching around the calculated range to get the transition.

SAGAN: Dave, if only ten per cent of the radiofrequency spectrum has been examined, then perhaps ten times (may be a little less) more molecules than you have listed are waiting to be found. If so, you will run out of small molecules. You will have to detect molecules bigger than the ones you have listed. You have properly described the history of this: even a molecule as simple as formaldehyde was thought to be bizarre and undesirable; but I think another line of evidence suggests substantially larger molecules are going to be found.

OWEN: The extent of the spectrum involved may not really tell you how many molecules are present. If we could just see more of the outer planet spectra we felt we could find more molecules, but that hasn't happened.

SAGAN: This seems to me very different. In the outer planets there are a few molecules which absolutely dominate the spectrum.

OWEN: There are reasons for that but I think there are similar reasons in David's spectra too.

SAGAN: But here we already see there are more than a few molecules.

BUHL: Some have been searched for and not found. Molecules derived from formaldehyde by the addition of a methyl group and so forth. That kind of a molecule may be down below present sensitivity levels. If we could increase the sensitivity by another factor of ten particularly at mm wavelengths we might find out a lot more.

USHER: How do you hope to increase the sensitivity?

BUHL: In the centimeter wave range the receivers are 10 to 100 times more sensitive, because of the use of parametric amplifiers and masers which are much more sensitive than the straight superheterodyne mixer receiver, which is what is used in the millimeter wave range. We have a factor of ten to one hundred to gain in receiver sensitivity, and probably another factor of ten in telescope area at mm wavelengths. Overall an increase in sensitivity of 10^3 is not an optimistic guess, so if there are big molecules with transitions in that region we can find them.

OWEN: Are all of these molecules in the galactic center?

BUHL: Yes. The galactic center is the largest molecule factory.

OWEN: If there is a source like that, whose frequency is very well known, you can collapse your box, and assign a much higher confidence level?

BUHL: Right. For example in Fig. 24 we show an Orion spectrum of hydrogen cyanide with the carbon^{12} and the carbon^{13} isotopes. Notice the velocity line up is good (less than 10 km/s). The center of these lines can be determined to within a few kilometers, so we are pretty sure that the scatter is not more than 5 km from the various molecules which are found in Orion. The X-ogen line was found when we were looking for the carbon-13 isotope, and we matched this line up in velocity with the known lines in Orion, and then worked backwards to get a rest frequency for this molecule.

Normally the problem is worked forwards. The rest frequency of the molecule is shifted by the amount of Doppler velocity that is known for the cloud. But in this case, working backwards, we had a known line with an unknown frequency, and we used the velocity to shift it back to an exact rest frequency ("Exact" means we could get it within a part in 10^5; whereas the box is more like a part in 10^3). Maybe it is academic to assign reliability since I agree a single line is not a perfect identification, but it would be difficult to defend putting another molecule in that slot. Most of the one line identifications won't stay that way for long.

Water is an annoying problem. We are looking for water through water, and so much of our atmosphere blanks out parts of the spectrum. The next transition up, at 180 GHz, is the only other microwave line of water. Thus to see this next transition up we have to look through about 40 db, which is about a factor of 10^4 attenuation from our own atmosphere. Some people thought about going to White Mountain or flying something of reasonable size (not a 36 ft. telescope) in NASA's Convair. We need to get up above a substantial amount of the water vapor to cut that 40 db down to 5 or 10 db. So water may remain a one line identification.

Most people have come away relatively unscathed from molecular claims. I haven't been wrong yet, but it's a very new field. Given a few more years, the debates and acrimony will develop, and we will begin to discuss reliability levels and so forth.

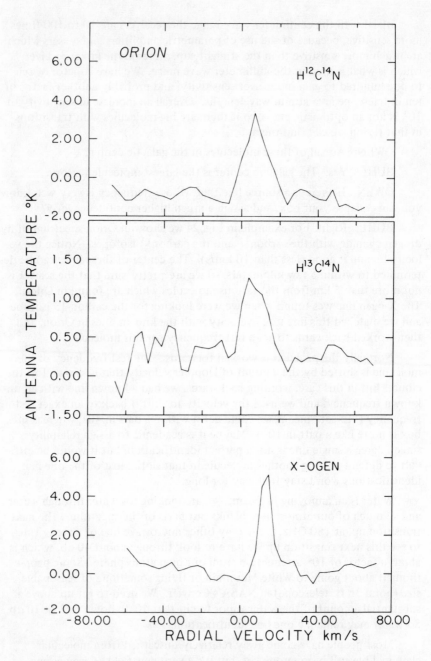

Figure 24. Orion spectrum of hydrogen cyanide with two isotopes of carbon, C^{12} (top), C^{13} (middle) and an unidentified line discovered during the search for $HC^{13}N$ (bottom).

SAGAN: This field, i.e., origin of life studies, has been through that same exercise!

MILLER: For the first time.

BUHL: Figure 25 shows the very intense water line in the source W-49. The temperature is about 2,000°K and the velocity is roughly +10 km per second. These features vary over weeks and months. Now the most intense feature is one of the triplets, whereas the main feature has diminished. The maser or whatever goes on in this little astronomical unit is apparently very turbulent and wherever the amplifying column is today may not be where it is tomorrow. It moves around. A very puzzling feature about the water clouds in this source is the presence of very high velocity lines. These go from -200 to +200 km/sec. Here is an example of a molecule filling up the box with a whole scatter of lines. How much high velocities occur in these very small objects is very strange, too. Water must be telling us something, but we don't know quite what yet. The masers are still open to considerable debate, but we think they are saying something about the energy transfer process in very early stages of collapse.

Table 13 shows rough values of relative molecular abundances. These vary by at least factors of ten, depending on who you talk to, so this chart shouldn't be taken too seriously, except that CO is much more abundant than these other molecules. The expected column comes from the known atomic abundances. Hydrogen is always higher by factors of 10^3 to 10^4, with carbon, nitrogen and oxygen roughly similar in abundance. Nitrogen is lower than oxygen by a factor of ten but for purposes of discussion they are essentially all about equal.

TABLE 13. RELATIVE MOLECULAR ABUNDANCES

Molecule	Observed	Expected
CO	10^4	1
OH	30	3×10^3
NH_3	30	3×10^2
HCN	10	10^{-1}
CN	10	10^{-1}
H_2CO	1	1

Cosmic Atomic Abundances

O/H	7×10^{-4}
C/H	3×10^{-4}
N/H	9×10^{-5}

The expected column is generated by multiplying these figures together and considering the random chance these atoms happen to get together to

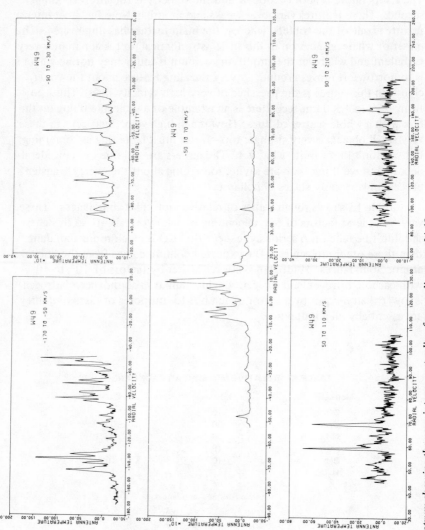

Figure 25. Spectrum showing the very intense water line from radiosource W-49.

produce a molecule. The differences come from the fact that some of these molecules are much more stable. CO, HCN and H_2CO are much better conditioned to the interstellar medium than some others such as NH_3 and OH.

Ammonia is very strange. So far it has been found only in the galactic center, despite the fact that we have four transitions here. Recently weak lines have been found in other sources but it should be a much stronger molecule. Nitrogen is tied up in other molecules known to be relatively widespread and abundant. Even if ammonia has some problem surviving, some of these others don't seem to be quite as afflicted.

MILLER: Isn't rate of synthesis as important as survival?

BUHL: Right. Some of these may have greater rates of synthesis. Although we have some calculations on survival we have practically no idea of the rates or even the mechanisms for generating these molecules, so that's even a blacker box.

The Goddard group has done some calculations on survival rates. The CO molecule survives longer than formaldehyde because its absorption bands are such that it can only be dissociated by very, very short wavelength UV.

SAGAN: Production rates may be very different, if the molecules are made out of littles, than if they were made out of bigs.

BUHL: Yes, if someone has ideas on generation mechanisms that might be taking place here, we might discuss them.

MILLER: To calculate the expected, do you just multiply the abundances and take the ratio of hydrogen or something?

BUHL: Yes, so these are normalized to formaldehyde.

FERRIS: Apparently some of these molecules have only been found from one or two sources. Are you implying that these are generally abundant in any large area, yet this abundance is what may be observed in one particular spot?

BUHL: Yes. The best value is OH to formaldehyde, where these two molecules have been measured in about fifteen or twenty sources. That ratio is roughly 30 to 1, whereas ammonia has only been reported in a few sources. So ammonia should be taken with a grain of salt.

Figure 26 shows another peculiar effect, the hyperfine splitting in formaldehyde, which has been used as another way of identifying the line. The arrows are the positions of the six hyperfine components.

ORGEL: What interaction are they due to?

BUHL: Nuclear spin interaction. The position of the components can be roughly seen by tracing the data points. The components seem to have relatively normal ratios, so it seems to be in reasonable equilibrium. This is an example of an absorption in one of the nearby dust clouds. What we're

seeing is an absorption against the 3°K background temperature without
any source behind to illuminate this formaldehyde cloud, which is very
peculiar. This three degrees worth of radiation still floating around is the
remnant of the fireball. No matter where in the sky the telescope is
pointed you will always see three degrees. The normal assumption is that
the molecule comes into equilibrium with the background radiation but in
this case it is cooled below three degrees. Apparently it is pumped down
to one degree, by collisions with the higher density molecular hydrogen in
these dust clouds, such that this can be considered the opposite of a maser.
The transition is being cooled down which causes an anomalous absorption
line.

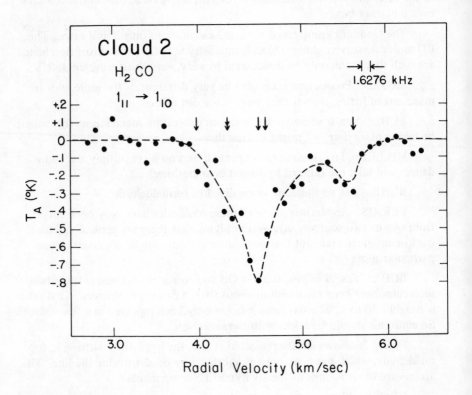

*Figure 26. Hyperfine splitting (arrows show 6 hyperfine components) of the formal-
dehyde line.*

I don't know whether this has ever been observed in the laboratory, but apparently formaldehyde does this under these particular conditions; this is our best explanation, which is all I have to say now.

YOUNG: Let's take a few minutes for coffee.

— — — — — — —

YOUNG: May we please reconvene?

Dave, have you more to say?

BUHL: A little, first about negative results. One of the most glaring examples is NO. There is no NO and this possibly has some biological significance in not being around.

OWEN: What about HS?

BUHL: That negative result is very difficult, because it is down in the middle of the FM band. Its frequency is around 100 megacycles so it was done at Arecibo (Puerto Rico).

The molecule most likely to come next from the Bell Labs group is H_2S which has a transition around 170 GHz. They have a receiver which operates in that range. In a month or two I expect we will have this.

SOFFEN: What about SO?

BUHL: Both NO and SO were looked for in the mm wave range but, as I mentioned, I am very uncomfortable about negative results. The molecule may not be there, or it may be there, but the transition may be in such an equilibrium that it can't be seen, i.e., it's transparent. With a negative result you are never sure if the molecule is absent or if the excitation conditions aren't quite right.

A very prominent example concerns formaldehyde in the Orion nebula, which is one of the strongest, most intense sources of molecules in the sky. The CO cloud up there is about as big as the moon. It is quite astounding. The moon put in the middle of the central star of the sword of Orion would give you a feeling for how big that CO source is. If you were sensitive to CO radiation rather than visible light, the CO cloud in Orion would be the brightest object in the sky. Yet when the early formaldehyde searches were done, formaldehyde was found in every source except Orion. Admittedly, at that point the only molecules known were those above H_2CO (Table 10, p. 228); we weren't aware of the more recent ones, but still it was very puzzling. Recently a 2 mm line has been found in the Orion nebula corresponding to formaldehyde, and it is the most intense 2 mm line of formaldehyde in the sky. But the 6 cm line is very weak, something like a tenth of a degree. The explanation, whether real or not, is as follows: The radio continuum, or hot region, is in front of the cold region. The molecules are behind and the ionized hydrogen and turned-on stars are in front so that because formaldehyde always appears in absorption there is nothing behind it to absorb. It is as though the light source were in front of the cell. Apparently this is why there was no appearance of formaldehyde, whereas the 2 mm line, which is an emission line, and does not require a background source appeared.

Several examples of this kind make me very uneasy about negative results. I certainly wouldn't use this to say that none of these molecules are present, because I'm sure they are, in some abundances. It is possible they will be picked up on other transitions.

FERRIS: How about CO_2?

BUHL: The problem with methane and CO_2 is that they are linear and have no microwave lines. Some, I believe, get bent in the excited state, but then you don't expect excited vibrational states. Yet it seems fairly obvious that methane must be there.   CH has been searched for, the problem is that there is not a good laboratory frequency for it. It is somewhere between 3 and 4 GHz, so people have gone up and down the spectrum looking for it, but lacking a precise frequency has made this kind of search real limited. Anyway we know from optical observations that it is there.

LEMMON: Do you find CH_2, a very important species in organic chemistry, methylene or carbane?

BUHL: We would like to. It should have some asymmetry to it. The problem is that these molecules are terrible to work with in the laboratory. Don Johnson has been trying to work with glycine and has yet to succeed in getting it in the gas phase long enough to be able to look at the spectrum. We are kind of stuck at that end of things.

Leslie mentioned CCH, an interesting fragment which we intend to look for. It is possible that it is close enough to X-ogen that we may be able to find it. We hope to look for these ions and radicals, essentially bits and chunks of larger molecules. If the chunks are there certainly by implication, the parent species would be expected to be present.

The Collapsing Cloud and Possible Interstellar Origin of Organics

BUHL: I'd like to discuss now the amount of organic material that might be in this cloud which condenses into a star, the problems of getting it into a planet, and the possible circuitous route of using a comet to bring a lot of material in at a much more hospitable state, when things have quieted down.

As Carl and Toby mentioned, a comet is a frozen sample of the very early solar nebula. We certainly know that impacts by meteorites are very frequent. In time scales of 10^6 to 10^8, there certainly are comet impacts. There was an event in Siberia in the late 19th or early 20th century which is thought to have been an impact of a comet. The probability of a comet impact in the earth's atmosphere is quite high. What happens to the comet as it comes down through the atmosphere and either volatilizes or synthesizes organic material cannot be precisely determined. In any case we have had contamination from the interstellar medium on the earth. Meteorites provide examples of very minute amounts of contamination, presumably comets provide much larger amounts. We might debate the significance of this source of organic matter.

SAGAN: The largest input is cometary debris called micrometeorites. The impact of the comet is less important than the stuff that rains down all the time from comets.

BUHL: Because the frequency is high.

SAGAN: Yes.

OWEN: But that may not be the most interesting with respect to these problems.

BUHL: What we would like is the mantle, not the core of the comet. It would be nice if the icy part came gliding in. Would you comment on how it would survive and in what form?

SAGAN: I don't have any objection to comets occasionally coming in to the earth.

SCHOPF: We appreciate that.

SAGAN: The question has to do with the contribution of such events to the origin of life. To summarize a long discussion the main point is there is evidence that objects like the earth and the moon melted at the final stages of accretion, due to gravitational potential energy which per unit mass goes as GM over R. But M goes as R^3 for constant density. Therefore, the gravitational potential energy goes as R^2. Thus at the final stages of accretion when R is largest, you have the greatest gravitational potential energy per unit mass. All your organics may be accreted cold but at the end they get fried. If the earth melted, much lower temperatures would still be high enough to pyrolize organic matter. Thus I'm very dubious about the cometary, nebular, and interstellar organic contribution at the very earliest stages of the formation of the earth.

BUHL: May I make one objection to that? It is true that the gravitational potential energy is the largest, but the heating goes as the rate of mass coming in, so it is also a question of how the distribution of accretion occurred. After the 10,000 years or however long it took for the hot earth to cool down again, one might still be sweeping out matter, but at a much lower rate.

SAGAN: Sure. During the first hundreds of millions of years after the formation of the earth, the tail end of the final phases of planet formation, there probably was still a lot of crud that all the planets were sweeping out, going around in the solar system.

But that is a minor contribution in terms of mass. Secondly, that material which would have been around in the solar system for a long time was not accreting cold. It would have been hot.

Thirdly, if this material fell on the earth before the earth outgassed then it wouldn't have persisted.

If the comet comes in, then the kinetic energy of the comet pyrolizes the stuff, yet if there is an atmosphere that permits some of the stuff to ablate and slow down as it comes in, then there would always have been an atmosphere of much larger mass than the comet brings in.

BUHL: I calculated this, and it seems if the comet had been big enough it could contribute a significant fraction of the atmosphere.

SAGAN: A comet that is massive enough can always be invented. Emmanuel Velikovsky drummed up a comet of the mass of the earth.

BUHL: No, I mean something 10 to 100 km, would have a significant amount of organic material.

ORGEL: How big are comets?

OWEN: Nobody really knows how big comets are, but people suggest diameters of 10, 20, 30 km.

SAGAN: A typical comet is a few kilometers, with a density of some tenths of a gram per cc, whatever mass that winds up to be. 10^{20} grams is a remarkably massive comet.

OWEN: For these days, as far as we know.

SAGAN: Yes.

BUHL: I have the number 10^{21} grams in the atmosphere. What is your number?

SAGAN: Roughly $4\pi r^2$ x 1000 gm per square centimeter – 10^{22} gm.

BUHL: If there is 10^{22} gms and 10^{20} is brought into a small area, this might be the trigger for setting something off.

SAGAN: Especially if it contained some special molecule that somehow couldn't be made here.

BUHL: I would like to invoke your shock synthesis as the best way. Do people believe in shock synthesis? [Laughter]

SAGAN: There are two schools of thought. [Laughter] There is no question that shock synthesis must exist. The worry of some is if amino acids could be made under the conditions as advertised (Bar-Nun, et al., 1970).

OWEN: Dave, would you please briefly describe your ideas about what occurs in one of these clouds after it collapses?

BUHL: Here are some rough numbers to think about (Table 12, p. 235).

I have already put up some of the densities. This is the environment before the collapse.

We assume this cloud is a mixture of gas and dust. The ratio of dust to gas is apparently about 10^{-12}, the number of dust particles divided by the number of hydrogen atoms. These clouds are primarily hydrogen, the hydrogen is up by about a factor of 10^3. There are vastly more amounts of hydrogen than these other molecules we are talking about.

ORGEL: Hydrogen atoms or molecules or a mixture of what?

BUHL: That is debatable. It may be roughly an equal amount of H and H_2. The H_2 detection is very recent. A problem of long standing has been how much hydrogen in the galaxy is tied up in molecular hydrogen. Only within the last year a measurement has been made with a rocket UV camera at about 1100 angstroms.

ORGEL: Did you say there were roughly 10^6 H atoms per cc.

BUHL: Per cc gets a little tough; there are about 10^2 per cc for hydrogen, and 10^{-2} to 10^{-4} for molecules per cc. This is terribly difficult to estimate because of the projection effect. You measure the average over the telescope beam, which may have lots of little, fluffy clouds in it, and you don't know how thick the whole mess is either.

A cloud is of the order of 10 or so light years in diameter. It probably fragments into several light year chunks, which collapse into individual stars, so star formation is a multiple process. It is not single event when the cloud reaches the stage at which it begins to contract. Presumably shock waves are involved in this fragmentation at an early stage of collapse.

During collapse the molecules, dust and everything are dragged along with the hydrogen. The maser may be the refrigerator that is helping it along, too, so that the molecule may be important in cooling the cloud during the collapse process. It collapses, and some of the angular momentum in the cloud becomes stored in the sun. At this point there is a problem of getting rid of the angular momentum and an easy way to do it is to blow away a lot of the gas used in formation and take the angular momentum out with the gas. There are a number of sticky unresolved problems with this model. The idea is that the cloud collapses into some sort of rotating disk object so that all the planets (except Venus which is peculiar) have the same direction of rotation. The angular momentum is all frozen into these objects.

The mass of a dust cloud with a density of 10^2 hydrogen molecules per cc and a diameter of 5 light years is about ten solar masses. As the cloud collapses 90 per cent of the hydrogen can be lost to get rid of the angular momentum. The details of this collapse process are not well known, but the dust, gas and molecules conglomerate. By a process of accretion planets and sun are built up. An asteroid sized mass will begin to have sufficient gravitational energy to "vacuum clean" the area out, but to get the accretion process going is very difficult.

Carl, do you want to comment?

SAGAN: Mine is not exactly the same problem.

In very diffuse interstellar space, characterized by the densities Dave has discussed, the abundance of carbon, nitrogen and oxygen is so small that assuming every atom that hits a grain 1000 Angstroms in diameter sticks, the amount of time before the mass of the grain doubles is several times 10^7 years! That is my objection to interstellar grains of that size. As the density goes up, particularly in a protosolar system, obviously the collision frequencies go up, and masses are accreted much faster. Then the sticking coefficient problem: how likely is it that an atom comes in and sticks? That depends on details of the solid state physics of the grain, both the chemistry and the detailed topology which nobody knows.

MILLER: It is a combination of coefficients. At higher temperatures they are about 0.1-1.0.

SAGAN: For grain atom collisions?

MILLER: No, these are a flat surface.

SAGAN: But that's right for a big grain and an atom.

MILLER: Wouldn't it be about the same thing at $3°K$.

SAGAN: This is not at $3°K$ but at higher temperatures. By the time the solar system is forming the dust cloud has already collapsed quite a bit.

ORGEL: Carl, are the energy quantities such that the difference (between atoms which stick chemically and those which do not) leads to fractionation?

SAGAN: Some people think so. Some think that we have an iron core in the earth because iron had a higher sticking coefficient than silicate, thus all the iron glops together before the earth formed, and then the silicates fell down on top. This gives you an idea that is very different than the views of core formation we spoke about earlier.

MARGULIS: Looking at the Orion cloud, is the theory good enough to project into the future with confidence and expect to see something much closer to a solar system when you come back?

BUHL: There are already at least two infrared objects in that dusty region where we believe are stars beginning to turn on, but these are very small objects. A size of one astronomical unit requires what is called very long base line interferometry which gives a resolution of a thousandth of a second of arc, which is at least three orders of magnitude better than the optical resolution. Essentially telescopes at different points on the earth look at the same source and get a resolution equivalent to a telescope the size of the earth, and so we have some idea of the size of the star, but it would be difficult to detect a planet at that distance. It's difficult enough to do it even on the very nearby stars.

Can you identify a planet in the orbit of a star by its motion?

SAGAN: Yes, for a planet of Jovian mass but only for the ten or fifteen closest stars. Even if the closest star had a planet of terrestrial mass, it would be indetectable by these techniques. There are other techniques, though.

Specifically to answer Lynn's question, the collapse time scales are very short. They tend to be millions or tens of millions of years.

ORGEL: Would electrostatic dusts with really strongly charged particles make it easier to accumulate or are they too small?

SAGAN: I don't know, Leslie. Some believe chondrules in chondrites originated by electrical discharges in the solar nebula.

ORGEL: But many of these carry with them substantial charges. It is not too clear to me whether the forces between the charged particles would contribute; but conceivably they might lead to very substantial nucleation.

SAGAN: It is an interesting idea which I don't think has been discussed.

LEMMON: David, although you probably are tired of chemists' suggestions of molecules to look for in the interstellar medium, I would like to mention briefly some favorites of mine. They are: carbane or CH_2.

Spectral details for this in the laboratory should be less difficult to obtain than others already mentioned. It is easily obtained in the photolysis of simple organic compounds. The methyl analog of any organic compound with a replaceable hydrogen can be gotten by a reaction with carbane.

BUHL: I have a vague impression that at the Ames meeting Zuckerman said something about CH_2.

LEMMON: I don't recall that.

BUHL: It is certainly something to look for.

The dust particles, at least ones we are reasonably aware of, presumably have an enormous size distribution. These are just the ones normally used in making models, but this doesn't mean there aren't grains even up to the size of the moon —

LEMMON: Some grain!

BUHL: . . . which have been able to accumulate. Once they get larger, they can get even larger, like the rich get richer.

LEMMON: It depends on how sticky they are.

BUHL: The radius discussed is in the micron region. The mass in the dust particles is roughly 10^{-3} of the mass of the hydrogen. Apparently quantities of dust are being discussed which might be comparable or even slightly larger than the molecules in terms of total mass. Possibly some kind of equilibrium goes on between the dust and the molecules.

BADA: Maybe this is my sheer ignorance, but what is interstellar dust?

BUHL: That is the key question which we have not been able to thrash out.

LEMMON: More or less the size is known, but the composition is not.

SAGAN: We really don't know the size of the dust. Some particles are in the size range of the wavelengths of visible light, because these grains preferentially scatter the light, but there may be much bigger ones. There also may be smaller ones.

BUHL: We assume this is an example of a large molecule because of the scattering and absorption well out in the infrared. The ideas of grains are obtained by indirect methods. We have not been able to measure anything spectroscopically except some indication of silicate bands.

ORGEL: Shouldn't metal ions be found in the dust?

SAGAN: Many metal ions are found in the interstellar medium.

ORGEL: Presumably ions will be involved in silicate if they are silicate. I don't, but at least some people have fairly good ideas of what ferrous and ferric ions look like in silicate media.

SAGAN: In fact, an identification of an iron rich silicate of a particular mineralogical structure in the interstellar medium has been made which some think is overly specific.

BADA: Couldn't it also be ice, isn't that correct?

BUHL: Yes. There were attempts to look for ice bands. This is an example of a negative result which I'm particularly hostile toward, because it caused some problems in the beginning when we wanted to search for water vapor. The Berkeley group did attempt to search unsuccessfully for a two micron band of ice.

OWEN: Hasn't Harold Johnson tentatively identified ice at 3.3 microns?

ORÓ: Leaving the humor aside, Dave, could you comment for those of us who are not astronomers on the diffuse bands revealed by Johnson's molecule?

BUHL: That is very difficult. There are about 25 interstellar diffuse bands which are somewhat broader lines seen in the absorption of a number of stars. They are interstellar because their shapes, their velocities, and everything are different. They are not associated particularly with the star, and have never been identified. They are in the optical range of the spectrum. Recently Fred Johnson came up with a porphyrin with a magnesium ion in the center which explains 16 out of the 25 bands, within some error.

SAGAN: He claims his molecule explains the 15 strongest unidentified interstellar diffuse optical frequency lines to plus or minus 2 Angstroms.

ORGEL: Is it true?

BUHL: Carl is in a better position to answer. He has probably thought about this more. Will you please give the rebuttal as to whether this is a valid identification?

SAGAN: First the molecule is a magnesium chelated tetrapyrrole, I will describe it although there must be a name for it: on each pyrrole there is a benzene on the outside.

ORGEL: Tetraphenyl?

SAGAN: There are two pyridine groups coming out of the magnesium at a funny angle.

MARGULIS: Has it a phytol chain on it too?

SAGAN: That's all that is missing!

ORGEL: How does it come to have 15 bands?

SAGAN: The Soret band explains the 4430 Angstrom strongest diffuse absorption — also this molecule cannot be free; it must be in a paraffin matrix in the interstellar medium.

Offhand, I would say this is a bit implausible. What happens if I replace hydrogen with an OH or if I exchange magnesium for iron? The frequency of all or of a significant fraction of the lines change by more than the observation will permit. This suggested precisely this molecule and no congener may be there.

USHER: No what?

SAGAN: No congener — that is, no slight variation on the molecule. There must be some beamed process which only synthesizes this and nothing remotely like it.

USHER: Has this been made on earth?

SAGAN: Yes.

MARGULIS: Has he done his own comparison spectrum?

ORGEL: Have you ever looked at the spectrum?

ORGEL: This molecule is very similar to one on which I once worked, and my impression was that they had one line at about 4,000.

SAGAN: But you didn't put it in a paraffin matrix, did you?

ORGEL: Then as I recall they have four further lines, close to 5500 or 6,000 if the molecules are unsymmetrical, and two if they are symmetrical, and not much else in the visible, which is what puzzles me.

SAGAN: Didn't you put the pyridines in?

ORGEL: The pyridines don't make much difference. The pyridines in shift the lines around, but doesn't change the number of lines.

YOUNG: Apparently Johnson is chairing a section on this work at the New York Academy of Sciences in June.

OWEN: Isn't he the same person who suggested chlorophyll?

SAGAN: Yes. This is the final evolution of that early statement. Three years ago with chlorophyll all 15 lines were not explained — just 4 were. He now has one molecule to explain everything.

BUHL: Lew Snyder and I have looked for pyridine, and somebody has yet to look for pyrrole. Unfortunately, it is in a gap in our range of available receivers. It's around 4.3 GHz, but that's one other check which certainly should be done. Of course, by no means is a positive result on pyrrole necessarily an indication that the porphyrin is also there. Having that many lines something complicated is suspected, but maybe not this horrible.

ORGEL: Why do you think it is one complex molecule?

MILLER: The real weakness of Johnson's story is his insistence that it be one compound.

BUHL: With larger molecules, are the lines more diffuse? What is your explanation?

SAGAN: They are diffuse in the astronomical sense, but apparently very narrow in the chemical sense.

ORGEL: I still prefer wave numbers. [Laughter]

SAGAN: What do you want to know? Dr. Usher, who has a slide rule, can act as my translator.

ORGEL: How wide is it?

SAGAN: Well, 4430 is about 20 or 30 angstroms, right?

OWEN: Yes. Correct.

SAGAN: Delta lambda over lambda equals delta nu over nu.

ORGEL: This is not very many waves, it is quite narrow by spectroscopy standards.

SAGAN: By our standard, very diffuse; by your standards, very narrow. And all of the others are much narrower.

ORGEL: 20 Angstroms?

SAGAN: 20 or 30.

ORGEL: That is too narrow for most big molecules.

SAGAN: Johnson says the problem is to make them narrower.

BUHL: Anyway, the problem we are stuck with right now is how to generate not only the prophyrin but some simple molecules. We can calculate relatively easily how long these molecules will hold together in the presence of UV fluxes. Louis Steif at Goddard has studied this (Steif *et al.*, 1972). Carl may have mentioned the numbers earlier; there is a 30 year survival time for most of these molecules, and maybe a factor of 10 to 100 longer for something like CO. These are survival times for a molecule in an unprotected environment. One of the reasons for wanting to put molecules inside of very dusty clouds is to prevent them from being destroyed by UV.

But what sources of energy required for the synthesis of molecules do we have available. If the synthesis is similar to that in prebiotic experiments, at some depth into the dust cloud we want to have UV radiation sufficient to generate molecules but not to break them apart later on. Again we have a very similar problem: how can the molecule be removed fast enough from the region where it is being generated not to be destroyed by the mechanism involved in creating it?

SAGAN: Several observers at the Ames meeting reported some of these small molecules in regions which lack dense aggregates. I call them clouds; and they seem to be more than the photodissociation travel time away from a possible source.

BUHL: CO is probably the most glaring example, however it has a much longer lifetime. The CO in Orion is 30 light years across but we are not sure the dust cloud is anywhere near that large. In the galactic center one can still detect CO 1000 light years from the center. We cannot see the center of our galaxy, so we don't know in great detail what it's like, but maybe the dust in the whole region is that large. It seems peculiar that CO is meandering around such a large region.

SAGAN: Similar remarks even about formaldehyde were made.

BUHL: Yes. Formaldehyde sometimes appears in the absorption away from dusty regions, but the optical data is rather incomplete. A large number of these sources are too far away to get optical data, since the nearby dust blocks our view of the more distant dust. This is the only way of knowing about the dust because the particles are too small to be detected by microwaves.

This question is up in the air at the moment. One possible suggestion that people are looking into is high latitude clouds. There are hydrogen clouds away from the galactic plane which are moving in at large velocities. There is some indication of molecular lines in these things and they are not obscured by the dust in the galactic plane.

FERRIS: Is the synthesis of formaldehyde improbable? Because if there is a lot of hydrogen and CO, this would be an obvious source for explaining a wide extent of formaldehyde.

BUHL: Stanley has some experimental things.

MILLER: We are investigating this, the problem is to make it at a temperature of $3°K$.

FERRIS: How about the numbers? Are there enough molecules of hydrogen and CO around together to come to the steady state concentration? Can you calculate that?

BUHL: Well, I think a third body is needed to take up the momentum.

ORGEL: An idea of the concentration of the molecules in space is 3×10^{-17} molar. Probably it still would be made in a place where the density is higher than average.

SAGAN: Yes. The critical problem about synthesizing organics is the lack of three body collisions.

ORGEL: Even two bodies are pretty unlikely.

SAGAN: Some of these certainly can not be produced by bimolecular collisions.

FERRIS: A problem with formaldehyde is that not just two but four atoms are hooked together and some of the energy is in vibration.

BUHL: If one is in an excited state, it can be done, but in a dust cloud it is hard to get one of these in an excited vibrational state. Polanyi has gone through a study of diatonics coming together and under what conditions you get combination. At the moment the grain theory is more satisfactory.

SANCHEZ: The probability is certainly higher. These grains have a great deal of material adsorbed on them, and then it's just a matter of collision with the grains. If it is UV energy or its equivalent that you require for synthesis, then, I guess, it would have to be on the outer edges of the cloud.

But I have heard speculations of shock waves as possible sources of energy in these clouds.

SAGAN: We are not restricted to ultraviolet sources.

SANCHEZ: What is a shock wave in a dust cloud?

SAGAN: It is rather different than the shock wave we are talking about. Most astronomical shock waves do not produce three-body collisions.

MILLER: Would you please explain the shock wave? If two dust clouds interpenetrate, do you get a shock wave? I don't understand that kind of shock wave.

SAGAN: A high temperature pulse moves through at some higher velocity and increases the pressure from nothing to six times nothing.

MILLER: But what temperature is associated with that?

SAGAN: A few thousand degrees, or so. It is instantaneous. It does not leave in its wake a set of radicals which can then be recombined by a three-body collision, as in a shock wave.

SANCHEZ: But can it result in reactions on the surface of the grain?

SAGAN: It can make radicals which weren't there before, and very slow recombination.

ORGEL: How hot can it heat a grain?

SAGAN: I don't know the answer to that.

MILLER: What is the method? Is it that if two grains collide, they get hot, or is there some other way of making them hot?

SAGAN: The shock wave is in the gas, not in the grains, but I haven't really thought about how a grain is heated with a shock wave.

BUHL: Stanley, the numbers that I took down from Ames were 10-20°K.

SAGAN: For the grains?

BUHL: These seem to be the numbers for the temperature of the grains. When you get into these clouds, the grain no longer radiates into space, it is surrounded by all the other grains, which are also radiating, giving an equivalent of infrared trapping.

SAGAN: Greenhouse.

BUHL: Energy radiated by one grain is collected by other grains, and vice versa, so these dense clouds do not return to the 3° equilibrium. The radiation only escapes on the outside edge of the cloud and is trapped in the interior.

MILLER: Could the grains get as hot as 70°K?

SAGAN: 77°K.

BUHL: With Carl's greenhouse effect a planet can be made as hot as the sun. It just depends on how dense these bodies are.

SAGAN: It is very interesting that we have molecules around which are easy to destroy and there doesn't seem to be an easy way to make them under general interstellar conditions. Under certain conditions we know exist these molecules would be easy to construct: namely, in a collapsing cloud. If produced in a collapsing cloud the synthesized molecules would be gravitationally bound and could not escape to the interstellar medium. Is it possible to tie these two events together? When a star is forming, and solar organic matter is around it, as soon as the star turns on, there is a huge amount of radiation and solar proton pressure which blows the solar nebula out. Observations of stars in this stage of formation, for example, T Tauri stars, show the dust blown back out again. This may be a very plausible contribution to the stellar grains, which implies carbonaceous chondrites have in them organics typical of the interstellar grain organic compounds. An attempt to produce the molecules in the interstellar medium by some fractionation process on the compounds in the carbonaceous chondrites might prove very interesting.

These events may be connected: molecules may be synthesized where it's dense, and then blown out to where it's not dense.

MILLER: How can they be blown out if they have a lifetime of only 20 or 30 years.

SAGAN: No. I'm suggesting grains, not molecules are blown out.

BUHL: But the grain, as Carl's calculation shows, has a very difficult time trying to accrete, an atom at a time. An alternate method is to suggest dirty stars, as Carl mentioned, which produce grains at a certain stage in their evolution and blow them out. The grains at this point are composed of some sort of carbon or silicate core, sufficiently well bonded that it can get far enough away from the star without being blown apart. Subsequently these grains accumulate mantles of ices and other organic debris similar to a scaled-down comet, in which the organic materials are stuck on the outside of this central core.

ORGEL: Wouldn't such grains be pulled back in again by gravitation? Could there be a model in which grains are formed way out, are perfectly stable, and gravitational forces then pull them in towards the center until a temperature sufficient to pyrolyze them is reached which produces the molecule? Would that work?

BUHL: At that point you probably have reached the stage Carl mentioned, where the gravity no longer allows the molecules to escape.

ORGEL: Above the gravitational field is where the temperature gets high enough?

SAGAN: Yes. It is really remarkable. By the time the densities are barely high enough to start interesting chemistry, we are already many orders of magnitude into a gravitational collapse.

BUHL: Once the grain has reached a certain distance from the star, it is no longer gravitationally bound, because of other objects to which it is equally attracted. The debris in the clouds is presumably the remnant of the previous generation of stars, and it is the raw material for the next generation. As the gas gets dirtier and dirtier, presumably the molecules get bigger and bigger.

ORÓ: How much work has been put into the process of the stars that either explode or put out solid materials? What reliability may we have in the published material?

BUHL: I can't give you a reliability factor, but many people are studying these early stages. The variable star phase occurs when a new star is proceeding along its Hyashi track onto the main sequence. Apparently there is sufficient material thrown out by these stars to account for dust that is seen. In any case, this seems more plausible than accreting the grain, molecule by molecule. In terms of reliability it is a more reliable method than accretion, but we don't know enough about accretion to be able to exclude it yet.

FERRIS: In this model, where do the molecules come from? Are the molecules blasted off the star?

SAGAN: By radiation. The material is chipped away by ultraviolet light, cosmic rays, and gamma rays. All kinds of things can do that.

BADA: Wouldn't that stuff be destroyed very fast though?

SAGAN: Rough calculations can be made but one is always safe for the reason Dave mentioned. Interstellar grains as large as you like can be invented and there is no special evidence against them.

ORGEL: I am a bit uncomfortable about that. As long as you find saturated molecules, such as NH_2CHO, it does not look as though they have been blasted off a surface, but rather have been chosen because they are stable. $NHCHO$ or NH_2CO can be made, neither would be stable, and they haven't been found.

SAGAN: Aside from that, suppose all three molecules are chipped off from the grains, one is going to have a longer lifetime in the interstellar radiation field than the others.

ORGEL: I question that. By "unstable" we usually mean unstable relative to reactions with other chemical agents.

SAGAN: Collisionally unstable, you mean?

ORGEL: Yes, rather than ultraviolet unstable, don't you think, Bob?

SANCHEZ: I'm not sure I follow.

ORGEL: The fact that NH_2CHO is seen and let us suppose neither $NHCHO$ or NH_2CO are, suggests there was a fair amount of time for ordinary chemistry to go on, rather than just a high energy or an accretion process from atoms.

BUHL: I would expect a fairly high density of radicals from bits and chunks. Already CO, OH, CN, CH have been found and I would not assume necessarily that these other molecules you mention are not there. They probably haven't been looked for.

ORGEL: If, in space there were, say, some NH_2 and a lot of CO and of course hydrogen atoms, wouldn't you think that if formation were just by collisions, only pairs of these together would have formed? Next you would expect to see CHO's and so forth, and would guess the next order of abundance would be molecules involving three of these things. I don't mean to make this point very strongly, but your list doesn't look as though they are determined this way. They look chemically fairly stable, in the conventional sense of chemical, for example, if you dropped them into water they wouldn't fall apart or do something peculiar.

OWEN: Maybe one should look for the fragments you mentioned.

ORGEL: Maybe they aren't there because they haven't been observed because nobody has ever seen them in the lab.

BUHL: On the model Carl has mentioned, where molecules are chipped off the mantle of some grain, often you may very well have a few dangling bonds.

ORGEL: Exactly. The observed don't, but probably molecules with dangling bonds wouldn't be recognized because they have never been worked with in the lab.

BUHL: We have some difficulty with them.

SAGAN: By "dangling bonds" do you mean free radicals?

ORGEL: Double radicals, triple radicals, and triplet states.

SAGAN: The frequencies are unknown of even amino acids, so certainly nothing can be said about many free radicals, which are harder to study experimentally.

SANCHEZ: Is it reasonable to look for an amino acid (which you mentioned) rather than a nitrile, or some other more volatile or earlier precursor?

BUHL: It is more reasonable to look for the precursors. At the moment we can't do anything with even something as simple as glycine.

SANCHEZ: The analogy is like looking for formic acid instead of hydrogen cyanide.

ORÓ: Where would you be looking, in an attempt to find glycine?

BUHL: We don't know what frequency to look at, so we're waiting for the NBS (National Bureau of Standards) group to come up with a frequency which can be looked at.

Some of these can be calculated. A simpler approach may be to look for bits and chunks like CCH or the amino and methyl groups of an amino acid, and maybe CH_2 to get an idea of how many of these incomplete molecules are lying around. By analogy if all of the bits of methane are there, certainly methane itself is there, even though there is no direct way of measuring methane.

OWEN: There are ways to detect methane, but not in the radio-frequency region.

ORGEL: What is happening in infrared radioastronomy?

OWEN: People are trying. It is difficult from the ground for the same kind of reasons Dave mentioned: places you would like to look at are often blanketed by water vapor.

USHER: I'm not going to be left with a very clear impression of the relevance of interstellar organic molecules to the origin of life on earth. Do you want to summarize the discussion about this, and tell us at what stage interstellar organic matter could have been relevant?

SAGAN: I feel the significance is analogical rather than material. The kind of processes that occur in the interstellar medium may have some analogies to the processes occurring in the origin of life. In general I doubt if there is a large contribution of interstellar organic matter to the origin of life on any planet.

USHER: When you were talking about the cooling of the earth, you said you were using a convection model, and then it would cool down in 10,000 years. How much convection did you allow?

SAGAN: The point of convection is that once some energy has been radiated away more energy is carried up.

USHER: To keep on going. But at some stage there must have been an exterior solid surface which is very, very hot.

SAGAN: It is a question of whether the earth freezes from the bottom up or the top down. Actually, it could be programmed from the bottom up. It's a question of the slope of the Clausins-Clapyeron equation, the PT diagram.

USHER: But, in your judgment, that the earth was hot at the beginning is not too much of a worry then?

SAGAN: No. We know that the heat input must have happened very fast, instantaneously, on the geological scale, but the amount of subsequent remelting is an important question.

One last remark: There is an experiment relevant to this interstellar problem, and perhaps to some prebiological organic chemistry on the earth as well. We select in various ways when we do these experiments. No one so far has selected for radiation stability. It would be very interesting to do an experiment in a radiation field for long periods of time, months or a year to see which molecules emerge, which are perfectly comfortable in a high ultraviolet environment. Even if there were certain molecules produced very infrequently, they may have dominated over a long period of irradiation. It would be nice to know which are stable. It would be nice to know if the molecule John Oró drew on the board is one.

ORÓ: Petroleum chemistry is actually affected by the stability of compounds analogous to this one. For example, vanadyl porphyrins are components in petroleum. I don't know the extent to which pyrroles are found in petroleum which would bring us to the initial question as to whether precursors to these would be found or not. But apparently it is a very stable molecule.

MILLER: Lunch is ready, but before we leave I would like to thank Dr. Shelesnyak and Mrs. Shelesnyak for the very nice accomodations, arrangements, and surroundings that added so much to the success of this conference.

REFERENCES

Anderson, D. L. 1971. The San Andreas Fault. *Sci. Amer.* 225:52-66.

Bada, J. L. 1971. The Kinetics of Decomposition and Racemization of Amino Acids in Natural Waters. In: *Nonequilibrium Systems in Natural Water Chemistry.* J. D. Hein, Ed. Advances in Chemistry Series 106: American Chemical Society, Washington, D.C. p. 309.

Bada, J. L. and S. L. Miller. 1968. Ammonium Ion Concentration in the Primitive Ocean. Science *159*:423.

Banda, P. W. and C. Ponnamperuma. 1971. Polypeptides from the Condensation of Amino Acid Adenylates. Space Life Sciences 3:54.

Barghoorn, E. S. and S. A. Tyler, 1965. Microorganisms from the Gunflint Chert. Science *147*:563-577.

Bar-Nun, A., N. Bar-Nun, S. Bauer, and C. Sagan. 1970. Shock Synthesis of Amino Acids in Simulated Primitive Environments. Science *168*:470.

Belton, M. J. S. and D. M. Hunten. 1968. A Search for O_2 on Mars and Venus: A Possible Detection of Oxygen in the Atmosphere of Mars. Astrophys. J. *153*:963.

Broecker, W. S. 1970. Man's Oxygen Resources. Science *168:*1537.

Cantor, C. R. and Tinoco, I. 1967. Calculated Optical Properties of 64 Trinucleoside Diphosphates. Biopolymers *5:*821.

Choughuley, A. S. V. and R. H. Lemmon. 1966. Production of Cysteic Acid, Taurine and Cystamine Under Primitive Earth Conditions. Nature *210*:628.

Degens, E. T. 1965. *Geochemistry of Sediments.* Prentice-Hall, Englewood Cliffs, N.J. p. 254-255.

Degens, E. T. and J. M. Hunt. 1964. Thermal Stability of Amino Compounds in Recent and Ancient Sediments, Humic Acids and Kerogen Concentrates. Intern. Meeting on Organic Processes in Geochemistry. Paris. Sept. 28-30, 1964.

Dekker, C. 1965. Separation of nucleoside mixtures on Dowex-1 (OH^-). J. Am. Chem. Soc. *87*, 4027.

Dicke, R. H. 1962. Implications for Cosmology of Stellar and Galactic Evolution Rates. Rev. Mod. Phys. *34:*110.

Doctor, V. M. and J. Oró. 1967. Non-enzymatic transamination of histidine with α-keto acids. Naturwiss. *55*:443.

Doctor, V. M. and J. Oró. 1969. Mechanism of non-enzymatic transamination reaction between histidine and α-oxoglutaric acid. Biochem. J. *112*:691.

Donn, W. L., B. O. Donn and W. G. Valentine. 1965. On the early history of the earth. Bull. Geol. Soc. Amer. 76: 287.

Florkin, M., Ch. Gregoire, S. Bricteux-Gregoire and E. Schoffeniels. 1961 Conchiolines de nacres fossiles. Compt. Rend. Acad. Sci. Paris *252*:440.

Folsome, C. E., J. Lawless, M. Romiez and C. Ponnamperuma. 1971. Heterocyclic Compounds Indigenous to the Murchison Meteorite. Nature *232*:108.

Folsome, C. E., J. G. Lawless, M. Romiez and C. Ponnamperuma. 1972. Heterocyclic Compounds from Carbonaceous Chondrites. Geochim Cosmochim. Acta (in press).

Fox, S. M. 1965. Simulated Natural Experiments in Spontaneous Organization of Morphological Units from Proteinoid. *The Origin of Prebiological Systems and Their Molecular Matrices.* Academic Press, New York.

Fox, S. M. 1970. In *Origins of Life I,* L. Margulis, ed. Gordon and Breach, N. Y.

Garrels, R. M., Mackenzie, F. T. and Siever, R. 1971. Sedimentary Cycling in Relation to the History of the Continents and Oceans. In: Robertson, E. D. ed. *The Nature of the Solid Earth,* p. 93-121. McGraw Hill, N. Y.

Gelpi, E., J. Han, D. W. Nooner and J. Oro. 1970. Organic Compounds in Meteorites III. Distribution and Identification of Aliphatic Hydrocarbons Produced by Open Flow Fischer-Tropsch Processes. Geochim. Cosmochim. Acta *34:*965.

Gibbs, R. J. 1967. The Geochemistry of the Amazon River System. Part 1. The Factors that Control the Salinity and the Composition and Concentration of the Suspended Solids. Bull. Geol. Soc. Amer., 78, 1203-1232.

Groth, W. 1937. Photochemische Untersuchungen in Schumann-Ultraviolet Nr. 4. Z. Phys. Chem. *B37*:315.

Groth, W. and H. von Weyssenhoff. 1957. Photochemische Bildung von Aminosauren aus Mischungen einfacher Gase. Naturwiss. *44*:510.

Groth, W. E. and H. von Weyssenhoff. 1960. Photochemical Formation of Organic Compounds from Mixtures of Simple Gases. Planet. Space Sci. *2*:79.

Hayatsu, R. 1964. Orgueil Meteorite: Organic Nitrogen Contents. Science 146:1291.

Hochstim, A. R. 1963. Hypersonic Chemosynthesis and Possible Formation of Organic Compounds from Impact of Meteorites on Water. Proc. Nat. Acad. Sci. *50*:200.

Hodgson, G. W., E. Peterni, K. Kvenvolden, E. Bunnenbey, B. Halpern, and C. Ponnamperuma. 1970. Search for Porphyrins in Lunar Dust. Science *167*:763.

Hodgson, G. W. and C. Ponnamperuma. 1968. Prebiotic Porphyrin Genesis: Porphyrins from Electric Discharge in Methane, Ammonia and Water Vapor. PNAS *59*:22.

Hoering, T. C. 1967. The Organic Chemistry of Precambrian Rocks, In: *Researches in Geochemistry,* Vol. 2, 87-111 (P. H. Abelson, ed.) John Wiley & Sons, N. Y.

Holland, H. D. 1972. *The Chemistry and Chemical Evolution of the Atmosphere and Oceans.* Wiley Interscience Press (in manuscript).

Holland, H. D. 1962. In: *Petrologic Studies: A Volume to Honor A. F. Buddington. Geologic Society of American.* N. Y., p. 447.

Hulett, H. R. 1969. Limitations on Prebiological Syntheses. J. Theoret. Biol. *24:*56.

Ibañez, J., A. P. Kimball and J. Oro. 1971a. The Effect of Imidazole, Cyanamide, and Polyornithine on the Condensation of Nucleotides in Aqueous Systems. In: *Chemical Evolution and the Origin of Life.* Eds. R. Buvet and C. Ponnamperuma, North-Holland Publishing Co., p. 171.

Ibañez, J. D., A. P. Kimball, and J. Oró. 1971b. Possible Prebiotic Condensation of Mononucleotides by Cyanamide. Science *173:*444.

Ibañez, J. D., A. P. Kimball, and J. Oró. 1971c. Condensation of Mononucleotides by Imidazole. J. Molec. Evolution. *1*:112.

Johnson, F. M. 1970. The Chemical Identification of Interstellar Porphyrin Molecule $C_{46}H_{30}MgN_6$ (χ). Bull. Am. Astr. Soc. *2:* 323.

Khare, B. and C. Sagan, 1971. Synthesis of Cystine in Simulated Primitive Atmospheres. Nature 232:557.

Kvenvolden, K. A., J. Lawless, K. Pering, E. Peterson, J. Flores, C. Ponnamperuma, I. R. Kaplan and C. Moore. 1970. Evidence for Extraterrestrial Amino-acids and Hydrocarbons in the Murchison Meteorite. Nature *228*:923.

Kvenvolden, K. A., J. G. Lawless and C. Ponnamperuma. 1971. Nonprotein Amino Acids in the Murchison Meteorite. Proc. Nat. Acad. Sci. *68:*486.

Lacey, J. C. and K. M. Pruitt. 1969. Origin of the Genetic Code. Nature *223*:799.

Lasaga, A. C. and H. D. Holland. 1971. Primordial Oil Slick. Science *174*:53.

Lawless, J. G., Kvenvolden, K. A., E. Peterson, and C. Ponnamperuma. 1971. Amino Acids Indigenous to the Murray Meteorite. Science *173*:626.

Margolis, J. N., R. A. J. Schorn, and L. D. G. Young, 1971. High Dispersion Spectroscopic Studies of Mars. V. A Search for Oxygen in the Atmosphere of Mars. Icarus *15*:197.

Margulis, L. ed., 1970. *Origins of Life,* Proceedings of the First Conference, May 21-24, 1967. Gordon and Breach, N.Y.

Margulis, L. ed., 1971, *Origins of Life,* Proceedings of the Second Conference, May 5-8, 1968. Cosmic Evolution, Abundance and Distribution of Biologically Important Elements. Gordon and Breach, N.Y.

Margulis, L. ed., 1972. *Origins of Life: Planetary Astronomy.* Proceedings of the Third Conference, February 26-March 1, 1970. Springer-Verlag, Heidelberg, Berlin and New York.

Miller, S. and Orgel, L. 1972. *Origin of Life.* Prentice-Hall, Englewood Cliffs, N. J.

Muljadi, D., A. M. Posner and Quirk, J. P. 1966. The Mechanism of Phosphate Adsorption by Kaolinite, Gibbsite and Pseudoboehmite. Jour. Soil Science *17*:212.

Oró, J., A. P. Kimball, R. Fritz and F. Master. 1959. Amino Acid Synthesis from Formaldehyde and Hydroxylamine. Arch. Biochem. Biophys. *85*:115.

Oró, J., D. W. Nooner, A. Zlatkis, S. A. Wikstrom and E. S. Barghoorn. 1965. Hydrocarbons of Biological Origin in Sediments about Two Billion Years Old. Science *148*: 77.

Owen, T. and C. Sagan. 1972. Minor constituents in planetary atmospheres: Ultraviolet Spectroscopy from the Orbiting Astronomical Observatory. Icarus (in press).

Paecht-Horowitz, M., J. Berger and A. Katchalsky. 1970. Prebiotic Synthesis of Polypeptides by Heterogeneous Precondensation of Amino-acid Adenylates. Nature *228*:636.

Pollack, J. B. 1969. A Nongray CO_2-H_2O Greenhouse Model of Venus. Icarus *10*:314.

Pongs, O. and P.O.P. Ts'o. 1969. Polymerization of 5'-Desoxyribonucleotides with β-Imidazolyl-4(5)-Propanoic Acid. Biochem. Biophys. Res. Commun. *36*:475.

Ponnamperuma, C. and P. Kirk. 1964. Synthesis of Deoxyadenosine under Simulated Primitive Earth Conditions. Nature 203:400.

Poroshin, K. T., T. D. Kosarenko and Shibnev. 1958. Isv. Akad. Nauk. USSR Otd. Khim. Nauk. 1129.

Rayleigh, Lord. 1939. Proceedings of the Royal Society A 170:451.

Reid, C. and L. E. Orgel. 1967. Synthesis of Sugars in Potentially Prebiotic Conditions. Nature *216*:455.

Report of the Geophysical Laboratory (P. H. Abelson, Director) in Carnegie Institution of Washington Year Book 69, p. 327. 1971.

Rossignol-Strick, M. and Elso S. Barghoorn 1971. Extraterrestrial Abiotic Organization of Organic Matter: The Hollow Spheres of the Orgeuil Meteorite. *Space Life Sciences* 3:89-107.

Rubey, W. W. 1951. Geologic History of Sea Water. An Attempt to State the Problem. Bull. Geol. Soc. Amer. *62*:1111.

Sagan, C. Jet Propulsion Laboratory Technical Rept. 32-34, 1960.

Sagan, C. and B. N. Khare. 1971. Long Wave-Length Ultraviolet Photoproduction of Amino Acids on the Primitive Earth. Science *173*:417.

Sagan, C. and S. L. Miller. 1960. Molecular Synthesis in Simulated Reducing Planetary Atmospheres. Astron. J. 65:499.

Sagan, C. and G. Mullen. 1972. Earth and Mars: Evolution of the Atmospheres and Surface Temperatures *Science*. 177:52:55.

Scott, W. M., V. E. Modzeleski, V. E. and B. Nagy. 1970. Pyrolysis of Early Pre-Cambrian Onverwacht Organic Matter ($> 3 \times 10^9$ yr. old). Nature 225:1129.

Siegel, S. M. and Guimarro, C. 1966. On the culture of a microorganism similar to the Precambrian microfossil. *Kakabekia umbellata* Barghoorn in NH_3-rich atmospheres. Proc. Nat. Acad. Sci. *55*:349.

Siegel, S. M., Roberts K., Nathan, H. and Daly, O. 1967. Living Relative of the Microfossil *Kakabekia.* Science *156*:1231.

Sillen, L. G. 1961. The Physical Chemistry of Sea Water. In: *Oceanography* M. Sears, ed. AAAS Publ. 67, 549.

Smith, J. W., J. W. Schopf and I. R. Kaplan. 1970. Extractable Organic Matter in Precambrian Cherts. Geochim. Cosmochim. Acta *34*:659.

Stawikowski, A. and J. L. Greenstein. 1964. The Isotope Ratio C^{12}/C^{13} in a Comet. Astrophys. J. 140:1280.

Steif, L. J., B. Donn, J. Glicker, S. P. Gentieu and Mentall, J. E. 1972. Photochemistry and Lifetimes of Interstellar Molecules. Astrophys. J. *171*:21.

Steinman, G. and D. H. Kenyon. 1970. *Biochemical Predestination.* McGraw Hill, N. Y.

Steinman, G. D., A. E. Smith and J. J. Silver. 1968. Synthesis of a Sulfur Containing Amino Acid Under Simulated Prebiotic Conditions. Science *159*:1108.

Stephen-Sherwood, E., J. Oró and A. P. Kimball, 1971. Thymine: A Possible Prebiotic Synthesis. Science *173*:446.

Tapiero, C. M. and J. Nagyvary. 1971. Prebiotic Formation of Cytidine Nucleotides. Nature 231:42.

Terenin, A. N. 1959. Photosynthesis in the Shortest Ultraviolet. Proc. First Intl. Symp. on the Origin of Life on Earth, ed. A. I. Oparin et al., Pergamon Press, N. Y. pp. 136-139.

Thompson, B. A., Harteck, P. and Reeves, R. R. 1963. Ultraviolet Absorption Coefficients of CO_2, CO, O_2, H_2O, N_2O, NH_3, NO, SO_2, and CH_4 Between 1850 and 4000Å. J. Geophysical Res. 68: 6431.

Usher, D. A. 1972. RNA double helix and the evolution of the 3′, 5′ linkage. Nature New Biology 235:207.

Van Hise, C. R. and C. K. Leith. 1911. The Geology of the Lake Superior Region. U. S. Geological Survey Monograph 52.

Van Hoeven, W., J. R. Maxwell and M. Calvin. 1969. Fatty Acids and Hydrocarbons as evidence of Life Processes in Ancient Sediments and Crude Oils. Geochim. Cosmochim. Acta *33*:877.

Van Valen, L. 1971. The History and Stability of Atmospheric Oxygen. Science *171*:439.

Weinreb, S., A. H. Barrett, M. L. Meeks and J. C. Henry. 1963. Radio Observations of OH in the Interstellar Medium. Nature *200*:829.